The Others, "The Whale People", describes with great integrity the author's life journey and the deep respect he has developed for some of the Earth's largest and oldest creatures. While others have written much about whales from a scientific viewpoint, few have been able to connect with them and glimpse into their being as Urmas Kaldveer has; fewer still have been able to express such experiences in writing. An engaging autobiography by a biologist with an inquiring mind and an open heart, readers of this book will reconsider the truths they believe are known about the consciousness of Earth's creatures.

Hillar Kalmar, translator Thirty Years a Mariner in The Far East, 1907-1937

The OTHERS is a record of the biological and philosophical impressions and reflections of a naturalist's journey in contact with wildlife of the large kind. One can easily connect with Kaldveer's' person through his rich life experience; unpretentious, yet filled with depth of view, as his writing draws us into his passionate journey, charged by the sea and the whales tantalizing us beneath the surface.

Richard Sears, Director, Mingan Island Cetacean Study Inc.

Provocative, entertaining, informative and well written, *The OTHERS*, "The Whale People", describes the experiences of a person with a great love and respect for all life and the natural world.

Urmas Kaldveer, "El Ballenero", mixes science, close encounters and dreams to show us his particular relationship with, and understanding of the whales, especially the humpback and blue whales. I was fortunate to meet Urmas some years ago and since our first talk I have always enjoyed his passion and energy when he describes his numerous encounters with "The Others".

Dr. Jorge Urban Ramirez, Director, Programa de Investigación de Mamíferos Marinos (PRIMMA) Universidad Autónoma de Baja California Sur (UABCS)

Also by Urmas Kaldveer

MAKIN' DO: A Single Father's Survival Guide to the Kitchen

CULTURES IN COLLISION: An Ethnohistory of the Huchnom

The OTHERS

"The Whale People"

A Personal Journey of Discovery, Transformation, and Healing

*To Mary-Beth
Welcome to the "Delphic Wave"!*

Urmas Kaldveer, PhD

BALBOA
PRESS
A DIVISION OF HAY HOUSE

Copyright © 2012 by Urmas Kaldveer, PhD.

All rights reserved. No part of this book may be used or reproduced by any means, graphic, electronic, or mechanical, including photocopying, recording, taping or by any information storage retrieval system without the written permission of the publisher except in the case of brief quotations embodied in critical articles and reviews.

Balboa Press books may be ordered through booksellers or by contacting:

Balboa Press
A Division of Hay House
1663 Liberty Drive
Bloomington, IN 47403
www.balboapress.com
1-(877) 407-4847

Because of the dynamic nature of the Internet, any web addresses or links contained in this book may have changed since publication and may no longer be valid. The views expressed in this work are solely those of the author and do not necessarily reflect the views of the publisher, and the publisher hereby disclaims any responsibility for them.

The author of this book does not dispense medical advice or prescribe the use of any technique as a form of treatment for physical, emotional, or medical problems without the advice of a physician, either directly or indirectly. The intent of the author is only to offer information of a general nature to help you in your quest for emotional and spiritual well-being. In the event you use any of the information in this book for yourself, which is your constitutional right, the author and the publisher assume no responsibility for your actions.

Any people depicted in stock imagery provided by Thinkstock are models, and such images are being used for illustrative purposes only.
Certain stock imagery © Thinkstock.

ISBN: 978-1-4525-5862-2 (sc)
ISBN: 978-1-4525-5863-9 (e)

Library of Congress Control Number: 2012917160

Printed in the United States of America

Balboa Press rev. date: 10/10/2012

Table of Contents

Chapter I	Early Impressions	1
Chapter II	A New Awareness	9
Chapter III	A "Call" to Return	20
Chapter IV	Ireland, Shamans and Whales	31
Chapter V	Life Changes and The ATOC Experiment	50
Chapter VI	Endings, Beginnings, Endings…	63
Chapter VII	SPLASH and The Baja Era	75
Chapter VIII	Clarity and purpose	96
Chapter IX	Implications	113
Chapter X	Discovery, Transformation and Healing	126

Appendix A Unforgettable Experiences 141

Premonition of Death, 1963 141
The White Wolf, 1990 142
A walk with Tomas, 1995 144
First close encounter with a gray whale, 2004 152
I see my first humpback in front of El Cardonal, 2004 153
My first close encounter with a mother and calf humpback, 2005 154
Blues and more blues, 2005 155
"Perseverance Furthers", kayaking with a mother and calf, 2007 157
In the midst of a "competitive group", 2007 160
Big day in the water with the "critters", 2008 161
Eyeball to eyeball with a blue whale, 2009 162
Swimming with a whale shark, 2010 163

Swimming with a pod of Orcas, 2011 165
Three close encounters in my kayak in one day, 2011 167
A memorable swim with sperm whales, 2012 169
Last swim with a humpback for the 2012 season? 171

Appendix B   A Short Biography 173
Appendix C   The Humpback Whale 177
Appendix D   Aboriginal Dreamtime 188
Appendix E   The Legend of the Golden Dolphin 192

In gratitude to "The Whale People" who reconnected me to The Great Mystery

THE LAST WHALE

When the waves are down and seas lie flat,
the humpbacked fossil stirs its sleep;
in grotto dark and ebb-tide mat,
this living sonar sounds the deep.

>An a cappella canticle,
>of endings and alone,
>a haunting ribbon melody
>that begs us to atone.

A song of sun and surface ride
(how easy love on waves was laid),
till ignorance and genocide
brought dread and fear of flensing blade:

Taught refuge in the quickened sea
and shadow-shrouded chasm walls,
till only in soliloquy
the last anachronism calls---

>In songs of searches in the deep
>(where lovers once had lain),
>a song of fathom fantasies,
>of love that might remain.

And dusty-rooted on the sand
I feel the contrapuntal flow,
that mammal mem'ry from the land,
I answer, "Yes, my friend, I know..."

I know my relic heart can bleed
in rhythm with your own,
and keen for lust and love complete,
...I too listen... all alone.

© Lyle Jan 1994
Rev Aug 200

Acknowledgements

Dr. Cadet Hand (deceased) who hired me at The UC Bodega Marine Laboratory and gave me the opportunity to test myself in the ocean, Ethan Silva for sharing a dream for a time, Scott Taylor for tweaking my scientific mode to include the spiritual, Thomas "Tomas" Pinkson for introducing me to the concept of "The Whale People", Richard Sears for reawakening my scientific interest in the whales, Vicente Lucero for being a perfect "piloto" and amigo for eight years, Louise Hay for the right words at the right time, Dr. Jorge Urban Ramirez for embracing my work and making me part of The Mexican Team during SPLASH, Hillar Kalmar for a valuable first read through, Alan Pomeroy for always being ready for the next adventure at sea and George "Jorge" Sievers for offering a helping hand to get this book started.

And always there to encourage, laugh and cry with me throughout the journey, my daughter Kersti and my son Zack whose loving presence in my life is beyond measure in value. We have all three shared "The Whale Experience", a dream that I hoped would someday be fulfilled....this year, 2012 it was. Thank you kiddos!

I would also like to acknowledge all those people who have donated to my research. Without your support I could not possibly have done the work, and my Mexican community here in El Cardonal who have so graciously accepted into their lives, the "loco gringo", who swims with the whales.

Urmas Kaldveer October, 2012

Foreword

For many years I have subscribed to an online discussion forum for serious marine mammal researchers. Daily, I read the postings, gleaning information about my own interests in this field. For me, as the Executive Director of the Cetacean Studies Institute, while the biology, ethology, and social dynamics of cetaceans are of interest, the main focus of my work lies in the area of human-cetacean interaction. In the mid-90s, an announcement in the forum describing a course being taught at the Mendocino Community College caught my eye: it seemed to stretch the bounds so prevalent on this scientist's information exchange site. Rather than a course focused entirely on biology, it seemed to be describing something more, something about how whales and dolphins have had an effect on human societies. Intrigued, I wrote to the author, off-list, asking him to describe in more detail what he was offering. The reply was friendly, and described a course that encompassed much more than standard fare. My enthusiastic response to this led to a long thread of correspondence, in which more and more details of our mutual fascination with the cetacean-human connection was shared. Thus began a friendship whose age is now approaching two decades.

At the time of this initial correspondence I was a focused, dedicated, and ambitious amateur, following an inspiration to open up a mystery. I had no academic credentials. I had spent two periods at University, some years before, with no degree to show for it. Not that I was defeated by an inability to understand or do the work required, but because life had other plans for me. The social unrest and paradigm shifts of the 60s and 70s, then marriage, fatherhood, and eventually divorce, all intervened. Along the way, a strange tale came knocking on my door, one that has intrigued me ever since. Called the Legend of the Golden Dolphin, it is a strange amalgam of mythology, science, spiritual teachings, ecological ideals, and a fascinatingly re-envisioned human history, in which dolphins play an important role. The Legend became my window into a 'parallel Universe', in which the interactions between cetaceans and humans hold special importance.

As our friendship developed, Urmas seemed quite willing to take my unorthodox views seriously. It was heartening for me to have an ally in a world to which I had very limited access. Scientists working with cetaceans, at that time, had little patience for anyone who was without University credentials. When Urmas described his work with Pelagikos, and the research voyages of Dariabar, I caught a glimpse of science in service to some of the mysterious goings-on that the Legend implied were of real importance. The friendly correspondence we established emboldened me at one point... I pestered Urmas to describe to me some of the inside details of the ATOC research the Dariabar was engaged in, and, without divulging any confidential information, he assured me that it was not as dangerous as the Animal Protectionist activists were claiming at the time. The ATOC experiments were very controversial then, an unknown that was feared by those who were working to protect whales. As you will read in this book, the ATOC work did seem to have the potential to deafen whales, or to cause damage to the sonic environment of the Pacific Ocean, and the activist 'industry' was on the case, raising fear-driven awareness of this potential disaster. I, too, felt a sense of outrage, that scientists could do something so dangerous to whales and dolphins, for what seemed little good purpose. Urmas responded to me in the midst of the experiment, despite the secrecy in which is was shrouded, assuring me that the people he was working with were of high integrity, showing a deep commitment to whale and dolphin protection, and that the experiments showed no negative effects for cetaceans. This was important information for me at that time, as I was able to write with a solid sense of reliable information to the activist community, in my efforts to calm some of the hysteria that prevailed (a brief summary of the ATOC experiment can be found at http://atoc.ucsd.edu/). I describe this episode to illuminate the open spirit in which Urmas conducts his life. It is meant to show his generosity, and his ability to hold science in the high regard it deserves, while understanding that the world beyond science is equally important.

Urmas and his wife came to visit me when I lived in Santa Fe. She was wary of visiting 'some stranger you met online', but Urmas and I found ourselves really enjoying our meeting. We seemed to find more and more in common as we spoke, sharing so much more than one can in emails. When I subsequently founded the Cetacean Studies Institute, a privately funded research and education project, Urmas graciously came aboard on the Advisory Board, adding much-needed academic credibility to the work. As the years have passed, we each have seen big changes in life circumstances. For me, a move to Australia meant a deeper dive into the Legend and the work I was doing, leading to a period when my wife and I designed and operated a wellness program, a kind of advanced dolphin-swim program, at a marine mammal rescue facility. We gave nine hours of classes to prepare people for two one-hour sessions, at dawn, swimming with dolphins. Sensuous, magical, quiet...then playful and full of delightful moments, the experience was transformative for many. After three years of this, the facility needed to undergo some changes and ended our program. This led me to return to Academia, where I am now, writing my doctoral thesis on Dolphin-Assisted Therapy. For Urmas... well, the changes are detailed in this fine book. He tells the tale better than anyone else could...

Urmas and I have stayed in touch, offering encouragement and occasional pieces of important information to each other, as we pursue the strange dreams that have led us to live lives dedicated to understanding the Others.

Over the years, I have found few academic writers willing to share, publicly, anything of their more personal insights into the mysteries surrounding cetaceans. Risking one's career, by appearing to be less than objective, and perhaps less 'factual', is not an appealing idea for most serious researchers. Teachers, however, do sometimes stretch beyond those limits, and Urmas is one of these brave people, both a scientist and a teacher, who understands that being inspired is the first requirement for good work, and being inspired means having one's spirit acknowledged. I have met, to be quite clear, many people doing scientifically rigorous work who do hold 'un-scientific' views as part of their personal worldview. However, few will openly share these views 'on the record'. In the book you hold in your hands is such a sharing, a kind of journal describing the path one man has taken that has led to a meaningful and satisfying life. And what more could one hope for? Being a small part of Urmas's story, sharing with him moments of delight, of change, of the growing peace that comes from having found meaningful work among cetaceans, is part of the Joy in my life.

This book is a rare gem, an entirely true story of a man's life that has had both trials and triumphs. A scientist, a teacher, a father, a soul in trouble, a man who makes seemingly strange choices, yet who finds healing and joy living beside the sea...this is a story to savour. Ripples from the passage of a whale are few: they leave behind no turbulence in their wake. We, as humans, aspire to do the same, although we do hope to leave a legacy. The story shared here, of a life that has succeeded in finding meaning, and healing, is an important one. It demonstrates the possibility of such a life, and if one man can do it, it can be done.

The Legend of the Golden Dolphin says the same. It offers a secret, a special key to open the mystery of a life, telling us that the living world of which we are part has always, and does now, hold the healing we seek. Among the 'Others', whether cetacean or other families of beings in this more-than-human world, are soul-saving connections to be made.

Urmas has opened a door for us all, one we can go through, into a world where science, spirit, the overcoming of fears, the healing of body and soul, and the Joy of communion with the 'Others' can be found.

Thanks, Urmas! I'll see you in the Whale Zone!
C. Scott Taylor, September 22, 2012, Australia

Preface

A little more than 50 million years ago a number of small and medium sized mammals began hunting near the ocean's edge. We are not sure what it was exactly that initiated this change in survival strategy for these four footed, warm blooded and furry creatures but it certainly must have owed it's success to an internal, genetic capability that had been "sparked" by a change in their environment.

Some believe that this spark may have been the much debated K-T event that decimated as much as 60% of terrestrial species and even more marine species a very long time ago. That event occurred perhaps sixty-five million years ago when a comet, or several pieces of a shattered comet, hit the earth in and around what is now The Yucatan Peninsula of Mexico.

The terrestrial mammals had begun their long road to recognition a good sixty million years before the K-T event but had yet to become more than a back drop, and a small one at that, to an earth dominated by The Dinosaurs. These occasionally very large but also chicken sized descendents of earlier reptiles, had successfully ruled the earth's global bioregions for one hundred and fifty million years by this time. They had done so, not because of their size, but because of their genetic ability to fill almost every niche (job) within the ecosystems they dominated. Indeed, had not the K-T event occurred it is quite possible that the reigning intelligence on our Planet would have evolved directly from these magnificent animals.

Be this as it may, something influenced these relatively small creatures, perhaps no more than a dozen different species, to begin a journey back to the marine world that is truly one of the great stories of animal evolution and has given us the great whales, the dolphins and porpoises, as well as other slightly less dramatic creatures like the seals, walruses, sea lions and sea otters. The fossil record shows clearly that by forty seven million years ago there was an air-breathing mammal, now named Ambulocetus, whose body would be recognized by anyone as being related to the great whales. Ten million years later the great toothed whales evolved and have not changed significantly for a good thirty-five million years. Today these early giants are represented by the sperm whale, the Orca, and the many different species of dolphin and porpoise. The baleen whales followed shortly after and are with us today as the humpback, blue, fin and a number of others.

Who are then these ancient creatures that chose to return to a home in the worlds oceans that their ancestors left three hundred and seventy five million years before? How did they re-adapt to this ocean environment and most important to the theme in this book, who are they now? Are they just very large, clearly benign animals, certainly with an interesting evolution, but otherwise without great distinction? How intelligent are they really? Does the song for example of the humpback whale mean anything other than an interesting phenomenon or is there more to these quite elaborate compositions? Do they recognize the concept of family or their own existence for that matter?

In this book I will try to answer some of these questions, but more importantly for me, it is the recording of the personal journey I have taken from pure research "Scientist", to friend and ally of The Cetacean Nation. This very intimate journey with the great whales, particularly the humpback and the blue whale, has prompted me to present my belief, not totally original but one that I came upon through intimate experience with many of the whale and dolphin species, serious scientific inquiry, and a good deal of meditation and contemplation. This is also the narrative of a very personal journey of healing that I believe has been mitigated by my intimate interaction with the whales and dolphins, particularly the humpback whales of The Sea of Cortez.

Sprinkled in among the chapters will be references to various periods in my life when either my physical, and/or emotional state, was in some disarray, and my journey, as may happen to any of us, had bogged down into depression and fear. I am quite certain that the trauma of my first three years of life in a war zone (Tallinn, 1941-43), and the atmosphere of numbing fear as my family escaped from Estonia across The Baltic Sea was instrumental in establishing a fear dominated personality. It was no more, or is now, anymore traumatic than that felt by millions of children in similar situations but it was enough to often hinder my willingness to participate in life's uncertain, but more adventuresome moments.

We humans lead very complex lives. In a world that sometimes seems beyond our understanding and often devoid of purpose, or we are physically/emotionally in need of comforting, we often seek the company of animals. It is my contention that among the animals that respond to that need most intently and thoughtfully, are the whales and dolphins.

In my own reality, the failure of three marriages and the sudden appearance of possible health issues broke my confidence and turned my life from one of moderate satisfaction into one of depression and fear. Although by this time I had been with the whales as director of Pelagikos, I had not yet opened myself up to The "Whale People" in a spiritual sense to the degree that I would allow any nurturing to take place. This changed when I began my life here in Baja Sur and found myself in the company of the humpbacks and blue whales that come here to breed and calve.

Even then it took a number of years for me to "surrender" to all the fears and failures, forgive my inner child, and allow "The Whale People" to begin their nurturing. I had also begun a daily routine of meditation, diet change and exercise, all of which supported the surrender. My main concern was cancer of which I was told I might have three types: prostate, kidney and skin. I made the conscious choice to disengage from western medicine, refused biopsies and chose to begin a self- healing process. All of this I carried with me while kayaking, diving and doing my scientific work as will be described in the text.

I would suggest that it has been in the evolution of a consciousness that has at its core, the desire to love and nourish other intelligent creatures that has dominated the world of the whales. What this implies is that all those anecdotal stories of positive, and sometimes life saving, interactions between humans and cetaceans may not be simply intuitive acts by "non-caring" animals.

What greater nurturing is there than to engage with another intelligent creature when the inner being of that creature is either in need of comfort or would like to play? I found myself entering the world of the whales with both a "heavy heart" and a desire to play and die. The "Whale People" embraced me and helped heal the person I was at that time. This book is the story of that journey. It is no more and no less.

My journey with these creatures began a long time ago while at San Jose State College working on my Bachelors degree in Zoology. That was 1961, it is now the summer of 2012, and at 70 years of age I find myself living with The Great Whales here on The East Cape of The Baja California Sur Peninsula in Mexico. January of 2012 was the beginning of my eighth consecutive breeding/calving season with the humpbacks of The Eastern Pacific.

Enjoy this personal journey of mine with these gentle and intelligent creatures, the ones I have come to recognize, honor and love....The Others, The "Whale People".

Chapter I

Early Impressions

Whales....big for sure, and that's about all I remember from my childhood and well into my adult life, just knowing they were big. Actually I didn't even have an inkling of what that meant. After all, I had never seen one, nor did I know anybody who had, but big they were, at least that's what everyone said. I didn't even know it was a mammal until I went to college. I assumed it was just an inordinately large fish. In that regard I wasn't really much different than anyone else of my time, this being the late '50's and early '60's. I suppose if I would have hunted through the literature, or maybe even spoken to a sailor (need not have been a whaler) I would have had a better idea of how extraordinarily magnificent these creatures were. Anyway, how does a person get to be near enough one to appreciate it, and then what? Yeah right... big, but hardly as interesting or certainly as accessible as so many other of Earth's creatures.

At a very young age I became deeply interested in animal life. Actually, I found all of the "life force" surrounding me as pretty extraordinary so I spent a great deal of time down at the creek (San Francisquito Creek) that flowed nearby our home in Redwood City, California.

Urmas Kaldveer, PhD

We didn't really live out in the country, but in the late '40's even the small thriving young residential area called Redwood City in Northern California still had a creek running through it. I spent hours in that creek, much to my mother's chagrin because I invariably came in contact with poison oak and suffered many a sleepless night in that special agony only poison oak can illicit. I guess it must have been worth it because I returned every spring and summer to investigate what new and exciting discovery could be made down there. I pretty much always went alone, most of my friends got bored easily and started goofing around, distracting me from my investigations. Most everything in and around the creek was small, little did I know that one day I would be just as close to the earth's largest creatures and swimming with them in their ocean world, as I was to the animals in the creek. What made these excursions to the creek so special was the profound peace that I found in being alone with this teeming arena of life, death, color, smell, shape and a million other things that made me wonder and marvel at the richness of what I eventually came to call, the "critters".

My days in the creek of course never evolved into a consciousness of the whales, but those days did instill in me a deep interest and adolescent, albeit unformed, love of our planetary critters. This naiveté regarding whales was to some degree lessened by reading the wonderful novel, Moby Dick, by that adventurer and author Herman Melville. This was required reading in my high school in Menlo Park, California, where we moved to in 1951. Menlo Park was a delightful, quiet town next to Stanford University and about 30 miles south of San Francisco in the west bay. Because of the proximity of Stanford University and the local developing electronic industries, Menlo-Atherton was an excellent high school, and I owe much of my early academic inspiration in biology to some of the teachers there, particularly Paul Castoro, my first real biology teacher. I didn't discuss Moby Dick with Mr. Castoro, I wish I had, but it was his love of all creatures that caused me to reflect a little more deeply on this Great White Whale that Melville wrote about. I didn't know until many years later that the story was more true than fictional. There was a large sperm whale that did ram and intentionally sink the whaling ship *Essex* off the South American coast in 1820. Only four men survived that incident to tell their story. Melville interviewed the first mate to get the background for the story that is known today worldwide and has prompted many a boy to go out to Sea.

Though I was fascinated by the idea of a creature so large that it could disable and then destroy a square rigged ship of 375 tons (I did know how big the ships were because my grandfather, a Sea Captain, took me to see the bark that is moored at The Marina in San Francisco) it was the interaction between the whaler Captain, Ahab, and the great whale, Moby Dick, that intrigued me most. How was it that an intelligent human being like Ahab, would see this "big fish" as such a personalized opponent and, give this whale credit for emotions not unlike our own? All of Ahab's relationship to this whale was in fact extremely personalized and it wasn't just because Moby Dick had taken his leg, it was about looking into the eye of Moby Dick and seeing the light of intelligence, and to Ahab malevolence.

Urmas Kaldveer, PhD

"to the last I grapple with thee; from hell's heart I stab at thee; for hate's sake I spit my last breath at thee."

—Moby-Dick, Ch. 135

It was this personalization of Moby Dick by Ahab that acted as a seed of curiosity, but only a seed. I did not search for more information to nourish this seed but as time went by the whales began to appear in my life and the seed took deeper root.

The fact was that there was little to be gleaned from the sources available to broaden my knowledge even if I had been interested enough to investigate further. This was after all the 1950's and interest in whales was only in regard to the valuable, but becoming increasingly less so, products that could be derived from a whale's body parts. The history of whaling for commercial interests has been long and inarguably colorful. Indeed, our human relationship to the whales goes back into the mists of time, long before recorded history, and is far more "colorful" and rich than I was aware. Later as an under graduate student at San Jose State College in 1964 I took a mammalogy course and there I was once again introduced to the great whales but now academically. During that semester we spent exactly 20 minutes in lecture and discussion on these creatures and then, since there was no individual to dissect, we left them. So in my reality the story remained the same, they are really big and we have been hunting them a long time. I still have the textbook from that class and have to laugh at the attempts made to depict them "scientifically" at that time. Certainly a clear indication of how little was known about them because the drawings are ludicrously inept. I graduated with a Bachelor's Degree in Zoology in 1965 and knew little more about these creatures than I did while roaming San Francisquito Creek in Redwood City thirteen years earlier. What is surprising is that the entire cadre of biologists, zoologists and mammalogists of the time, knew very little more than I did....and that is astonishing.

Two stories regarding human relationship to the great whales began to emerge in the 1960's. One that everyone pretty much is aware of if for no other reason than Melville's book and the movies that have been made from it. That story being our interaction with the whales through "commercial whaling" as stated before, and it being a limited interaction to say the least, knowing what we now know of their extraordinary intelligence and natural history. This is the story we are all most familiar with and it is one that should be examined, even if briefly. The other story is the one that comes to us through the experiences of our more ancient ancestors and the men and women who have chosen in modern times to devote their lives to understanding just what, and who, these creatures really are? Ironic isn't it that maybe Ahab was on to something. I will delve more into this other story in the next chapter.

So, when did whaling begin? Why it began is clear, it was for food. Not luxury food, just food for survival. To be sure it was not commercialized at this point for indeed we are talking about a period beginning a good six thousand years ago when most of the worlds people did not even possess a written language. Nevertheless, there are petroglyphs depicting whale hunting in the Baltic Sea that may date to even earlier times, perhaps as far back as eight or ten thousand years before the present.

The Arctic Circle boat people (including the Baltic Cultures) who developed whale hunting, not only created large boats, but also their quest for whales took them far into the sea as they searched for them. It is speculated by some, that these same people, due to their long range capabilities and their willingness to sail in very rough seas may have been the ancestors of the colonizers of The Atlantic Coastlines and the progenitors of the "Megalithic Cultures" of southern Europe.

Later, various people throughout the world began hunting whales and dolphins for food as soon as they realized that a whales or dolphin's body could provide any number of nutritious and otherwise valuable products to help sustain the hunter's lives. It was not necessary to kill a great many because as hunter-gatherers their populations were relatively small, say a maximum of 50-60 individuals within a settlement. It was also very risky to try for a kill on the larger, healthier whales so the hunters focused on the smaller, slower, weaker ones: not unlike other planetary predators. Evolutionarily it does not benefit a predator species to put more calories into a hunt than can be derived from the kill. The results were that the numbers of whales were never seriously diminished and the hunters culled the weaker within the species, thereby insuring the strongest genes within their prey.

The first mention of more modern whaling begins with The Basque People and was made in 1059, when it was said to have been practiced at the Basque town of Bayonne. From there it spread to a number of European countries and by the 1600's had become commercialized.

The balance then between predator and prey no longer applied and by the 1800's whale populations worldwide were beginning to show signs of degradation. During the 1700's the world's populations had begun to burgeon and it was discovered that whale oil made a very fine, non-sooty flame and was also such a pure oil that it could be used as a lubricant for the most delicate instruments. It was not long before the variety of products that could come from a whale's body were so vast and of such high quality that it became "commercially" feasible to hunt them at long distance. With people to supply, ships to sail, and men more than ready to profit, the balance between predator and prey came to an end and widespread hunting lead to significant declines in various whale species numbers.

Here is a partial list of products that have been derived from the bodies of whales; in all there are over 300.

Lamp Oil, Corset Stays, Linoleum, Soap, Margarine, Paint, Dynamite, Transmission Fluid, watch Oil, Buggy Whips, Golf bags, Buttons, Fertilizer, Cattle Food, Fuel, Umbrellas, Tennis Racket Strings, Drum Heads, Insecticides, Varnish, Piano Keys, Fishing Rods, Car Wax, Shoe Polish, Iodine, Insulin, Carbon Paper, Crayons, Bar Stool Covers.... etc.

By the 19th and 20th centuries whaling had become a very lucrative industry and practiced by any nation that could afford the ships necessary to hunt them. Even without an ocean going fleet of whaling ships it was possible to do coastal whaling from land stations located along the migration routes of the whales. Most all whale species swim relatively close to shore when on these migrations and once the routes were established it was also possible to follow them to their breeding/calving grounds and their summer feeding grounds. With the advent of the faster steamship and the invention of the exploding harpoon even the deeper water and larger whales like the blue and fin were no longer safe.

Throughout the early 20th century the whaling industry continued to prosper but there was less demand for the products, mineral oil had replaced whale oil as a cheap and clean lubricant and the petroleum industry was turning out similar products cheaper than could be produced from whales. On top of that, the whale populations all over the world had been so decimated that many species were simply no longer in numbers large enough to warrant the expense of finding and killing them.

During all this time, little or no effort was made to study the whales, but by the 1960's a new breed of scientist was in the field and what they discovered demanded a new awareness in regard to our relationship to the whales and marine mammals in general. I too was about to become a (small) part of that early recognition of this "Other" sentient planetary creature that would eventually come to astound us in their complexity and intelligence.

Chapter II

A New Awareness

In 1969 I graduated from The University of Arizona, Tucson, with a Master's Degree in Medical Microbiology. I continued there for an additional two years working on my PhD in Immunology, which at that time was a very "sexy" field of study. Due to a changing political climate, some experimentation with various drugs and a predilection for dirt biking in the desert rather than working in the lab, I chose to leave the University and I headed back to The Bay Area in California to discover just exactly what it was I really wanted to do. With me now was my wife of five years and an infant daughter, Kersti, who changed my life the way children can do. She is now forty three and one of my best friends as is her brother Zack.

Feeling a need to escape academia for a while and the intensity of "The Counter Culture Movement" within a University setting, I decided to see if I could get a job working for P.G. & E. (a large California energy company) as a laborer. I had worked for them every summer as an apprentice lineman during my undergraduate years at San Jose State College so I figured a job would be in the offing. I didn't want to be a "hiker" anymore so I was hired as a "ditch digger" (the commonest of laborers) and was assigned to a gas crew laying services to residencies in the San Francisco bay area. It was a wonderful hiatus from the intensity of grad school, new fatherhood, political activism, and it was simply a good way to rest my brain. I enjoyed every minute of it and was always on call during lunch to tell stories in between hands of Pedro (a very short version of pinochle) about higher education, my travels in Europe and what my crew thought was a strange but exciting and different life.

While living in Palo Alto, California at this time, my wife and I decided to take a drive up the coast to Bodega Bay for a day by the sea. I had no particular affinity for the ocean, in fact I had a deep, subconscious fear of it which included a reoccurring dream of being in the sea (no boat) in a great storm with very dark skies and menacing waves and seeing a particularly threatening wave cresting as it nears me. Nevertheless, the California north coast is an incredibly beautiful part of this planet so a visit was something to look forward to and enjoy. As we drove around the north end of the harbor there in the town of Bodega Bay I saw a large sign announcing the location of The University of California Bodega Marine Laboratory (UCBML).

I stopped and jotted down the telephone number (the gate was closed and locked) and the address. With images of Jacques-Yves Cousteau running through my mind (he was after all the popular "Godfather" of oceanic research) and the dramatic and exciting work he was doing, I decided to write the laboratory a letter of introduction and apply for a job.

I had read a fair amount of Cousteau's work and his thoughts on everything from ocean pollution to whales, and I was fascinated! I thought may not be in the depths of it like Cousteau at UCBML but at least I would be near the action, and after all, I have a terrific imagination. Well, The Gods had their own agenda for Urmas Kaldveer. I intended to seek a position as a microbiologist and that was what I inquired about in my letter. I knew nothing about marine microbes but to me a bacteria was a bacteria and not much more (this misconception would be dramatically and delightfully corrected by Lynn Margolis' book, *Microcosmos,* many years later). They said no, but they were looking to hire a specimen's collector! Suddenly the Cousteau fantasy began to loom larger as a possibility on my occupational horizon and far closer to what I had imagined in my somewhat unrealistic fantasies. The problem was that I still had a significant fear of the ocean... "oh well", as they say, I wouldn't let that deter me from what could be a once in a lifetime opportunity. I had to, well, not lie exactly, but not really be forthcoming when answering their questions about experience (later to get me into serious trouble more than once, not with them, but with the ocean). When asked about my SCUBA diving experience I replied that yes, I was certified. I was but I had only made one ocean dive! When they asked about my boat handling experience I answered that I had some (my brother and I had rented an outboard on Lake Merced in The Sierras once to troll for trout), so you see, I was not exactly forthcoming.

Nevertheless, in 1970 I actually got the job. Many experiences followed that deceit on my part, mostly positive but some, as implied, not so good. I certainly did not live up to the trust my employers had extended me, nor did I come close to emulating my hero, Cousteau. What did, however, come from those two years at UCBML was a new appreciation for the ocean and the beginnings of a direct and intimate connection with the great whales. The connections were to prove to be in the realm of the personal/spiritual rather than in the academic since no one there was engaged in marine mammal work at the time and therefore there were no opportunities on my part to be near any "whale action" in a scientific sense. My experiences with whales during those two years were purely unexpected and personal.

There were only two experiences really, but both literally changed my awareness of this planet that I lived on, and sparked what would later become the love and devotion I have for these wonderful creatures. In a very real sense, they also initiated the work I have engaged in since the 1990's...but this would however be a long time after my Bodega experiences.

The first experience I will relate is the one that was, and continues to be at the core of my spiritual connection to the great whales, especially the humpback whale, Megaptera novaeangliae, that is the focus of my current studies (see Appendix E).

One rather slow day at the lab I received a phone call from one of the professors at Berkeley needing some specimens of a particular fish available in the outer Bodega Bay area. I had begun noting where the best places were to acquire popular research specimens so I was confidant that I could fill the order. I said I would get them to him within the next few days. It was still early in the morning so I went around the lab to find someone to go out with me to help with the lines. As it turned out no one was available. I really did not want to go out alone but I had promised the professor so I got my gear together and took *Sulu Girl*, our 35 ft cruiser, out through the channel and into the open ocean.

This is a picture taken in 1970 of me preparing for a collection run on the Bodega Marine Laboratory boat, *Sulu Girl*. I usually went out with at least one assistant to help with the collections.

I noticed that there was a thick fog bank a ways out but it was hard to tell how close and how fast it was moving, but I figured I had plenty of time. Once I got my lines out I rechecked the fog bank and yes, it had come in rapidly and from not far away. I hauled in my lines because I did not want to be caught in a fog so close to the cliffs with a significant swell running (the north coast of California is historically notorious for sunken boats and ships that have gone up on the rocks of that craggily coast). I did not have radar or sonar so I had to do my navigation by depth finder and chart. Not being particularly efficient at this sort of thing, I realized that there was indeed a real good chance I would end up on the rocks. If this was the case I was very definitely in serious danger, perhaps more than I had ever been in my thirty years of life so far.

The fog was miserably thick, I could barely see the bow of the boat and I felt as alone as I could possibly imagine. My knees were actually weak! Suddenly I heard a loud eruption of water in front of me and a huge black mass exploded from the sea, arched in front of the boat and came down with a thunderous splash, rocking *Sulu Girl* and just about stopping my heart. I knew it was a whale (the first one I had ever seen) but I had absolutely no idea what else to think except, "damn, that was close" and "I hope it's friendly".

I couldn't see it but it started to circle my boat, blowing loudly and not causing a ripple, but I knew it was close, real close. This went on for an indeterminate time during which my attention was drawn away from the possible disaster, and I begin to feel the energy of this whale, and it gave me strength and hope....I am not alone.

When the fog cleared, I found myself directly in front of the jetty leading into the harbor, a safe place compared to where I had set and hauled my lines. The whale left at the moment the fog lifted and I didn't see its blow again. I contemplated the significance of this event for quite some time until I came to the conclusion (reinforced now by dozens of similar but not as dramatic moments) that this whale knew I was an intelligent creature like itself and decided to help me because it also sensed that I was in a dangerous situation. I have had a number of people hear this story and they suggest alternative explanations. It does not matter, I know that that whale could feel my desperation and decided to help a fellow intelligent creature. Did this experience compel me to start on my journey with the whales...oddly enough, no. I did not see another whale for twenty years and than it was they that called me back to the sea.

Before I tell you about that, I want to tell one more story regarding that period at UCBML and whales.

After my personal "whale experience" as described above, I can't really say that I fully understood, absorbed or assimilated the ultimate influence it would have in my life but I can say that I felt different. This was most actively manifested in an interest in reading more about the cetacean world. One of the books I read included some information regarding a young man who had designed a set of metal percussion instruments and used them to attract whales. I was later to meet him and we had a discussion about my experience and others that he had had or heard about. The one that struck me deeply and became a whale tale that I told for years after, was when he was asked by a young couple who had just lost their child if he would call in the whales and then scatter the boy's ashes on the water near them. When they asked my friend this, he had to tell them that it was not the season in that area for whales to be in the vicinity so they would have to wait. They were adamant however and he finally agreed to attempt it.

He went out in a boat and then placed himself in the water in a converted lawn chair with pontoons and the drums in front of him. On the surface of one of the drums he had pasted the picture of the boy who had died. As he drifted away from the boat he began to play his aquatic instrument. He was quite sure it would be to no avail, yet he continued to play with a deep commitment to these grieving parents. The parents were in the boat also sending energy to the ocean in hopes of aiding my friend. Suddenly, and literally out of nowhere, two gray whales surfaced on either side of him and close enough to almost touch. At first he was a bit frightened but continued to play and then released the ashes on both sides of himself. The whales stayed a little longer and then just as suddenly glided deeper and were gone. My friend told me that it was one of the most enlightening experiences he had ever had.

These two experiences (one vicarious) shifted but did not alter my own designs at the time as to what I would do in my future. I left the marine lab in 1972 and was not to see another whale for twenty years.

In September of 1973 I began a career as a Community College instructor and then professor. I taught pretty much everything in The Natural Sciences, some chemistry and even some history. While I was teaching at the college, major changes were taking place in our understanding of the whales and dolphins. Not the least of which was the discovery that our human relationship had not in fact always been one of hunting and exploitation of the cetaceans. Indeed, archeological evidence indicated that prior to the advent of large scale commercial whaling in the 18th century, and even before the balanced whale hunting of the early Baltic Peoples, various cultures throughout the world had developed intimate spiritual and celebratory relationships with these marine mammals. The people who manifested this perhaps more than any other culture, and still do, were certain coastal clans of The Australian Aborigines. The Australian Aborigines of today are the direct descendants of the first group of Cro-Magnon people to leave East Central Africa and venture from our common home in that continent into what would later become all of "The Old World". They were a coastal migratory people and made the journey from Africa to Australia in less than 15,000 years. Their relationship to the ocean waters brought them into constant contact with the cetaceans who also dwelled along the coast and still do. Rather than hunt them, they found something of a kinship with these animals and began to weave them into their life ways and their mythologies (see Appendix D).

They were not the only ones. Recent studies of the early pre-Hindu cultures, the archaic Middle Eastern Cultures, African, early Greeks and later Romans and all the way across the Pacific to the Native Americans in "The New World", a sense of kinship had developed between cetaceans and humans. My friend Scott Taylor refers to this in his outstanding history of this contact (*Souls in the Sea*) as "The Delphic Wave". A wave of awareness passing through culture after culture, generation after generation, manifested in celebrations, stories, myths, dances and even worship. "The Whale People" were seen to be soulful, intelligent and benign. Little attention was given to this early human/cetacean connection by the scientific community during the modern age. Nevertheless, a new breed of scientist was emerging in the 1960's and was rediscovering The Delphic Wave through new tools.

The ancient legends, myths and "Whale Lore" of those long ago times are still etched and painted on the walls of ancient ruins and found in many of the earliest texts regarding observations of the natural world. Rather than reiterate the long and fascinating saga here, I will enthusiastically refer the reader to Scott's book, his eloquence in telling this story is beyond my ken to emulate.

To some extent it was the growing return of The Delphic Wave that inspired a number of the new breed of scientist during the 60's to engage with the cetaceans on a level unprecedented in human history. With the tools of science they were able to reconnect with the cetaceans but this time both academically as well as spiritually. Early in this period Roger Payne began his studies of humpback whale song, Lewis Herman focused on communication and learning abilities of dolphins, John Calambokidis began his ID work to establish numbers and migration patterns, Richard Sears became devoted to the blue whales with the same intent, and then there was John Lilly.

John Lilly was the first of these early scientists to suggest the possibility of a significantly higher level of intelligence in these creatures than even his colleagues were willing to admit to. Working intimately with dolphins, Dr. Lilly came to the conclusion that they were not only intelligent but exhibited, and indeed were possessed with higher intelligence, meaning a level of cognizance equal to our own. He was of course relegated to the weird category of scientist for these beliefs but nevertheless his book, *The Center of the Cyclone*, published in 1972, became a primer and inspiration for an entirely more sophisticated approach to what he called "The Cetacean Nation". Though his ideas were often ridiculed he devoted the last years of his life to the study of consciousness within the animals he had come to respect and love. As a graduate in physics and biology from The California Institute of Technology, his work could hardly be disregarded. Unfortunately his work was seen, as is so often the case (a good example is Nikola Tesla), as a threat to the current paradigm; he died in 2001 without the full appreciation within the scientific community of his findings. I did read his book and was enthralled, but my life at that time was far busier than I had anticipated. Now I had full custody of two children, a broken heart and some serious doubts as to the proper direction my country was taking.

While all this was taking place, I was passionately engaged in a teaching profession that I came to love and that became my salvation. It would be some time before the work of these men would affect my life, but affect it, it would.

•

Chapter III

A "Call" to Return

By 1987 I had remarried, divorced again six years later, my daughter was off to college and my son would graduate from high school in three more years. I had taught Biology, Chemistry, Anatomy, Ecology, Environmental Science, Field Biology, Microbiology, California History, Russian History, Contemporary U.S. History and... Marine Biology. It was in teaching that marine biology course at the college in 1989 that I was once again made aware of the marine mammals and particularly the great whales. I lectured about them but never saw any, nor felt compelled to engage in a devoted literature search to broaden my perspectives about them. Nevertheless, I was reminded of my earlier experience at Bodega Bay and indeed wondered what it might be like to actually be involved in fieldwork on the ocean. My fear of the ocean had dissipated somewhat due to the work at Bodega and my fantasies around a Jacques Cousteau like avocation still lingered in my mind as a kind of "pipe dream".

In 1989 I remarried yet again, number three, and found myself visiting Mexico a number of times with my new wife- due to her love of warmth, the sea and the culture of Mexico. For that reason alone I am indebted to her, for she introduced me to the culture I would adopt as my own twenty years later. I of course did not know this or frankly even imagined it but apparently the whales had an agenda for me that I was simply slow to perceive.

In 1990 we took a vacation to the then small Pacific Coastal town of Zijuatanejo, Mexico. We were told there was a secluded and pristine beach a short hike from the south end of the town and we decided to investigate it. Well, it was for certain the beach of one's dreams. We had brought our snorkeling gear and took full advantage of the beautiful near shore sea life that was there.

After one of our forays into the little cove fronting the beach I returned to shore before Susan and just sat gazing out to sea in this magical place...and then I heard a voice! It was a deep resonant voice but I couldn't tell where it was coming from. It seemed to be coming from beyond the cove, somewhere further out at sea. I shook my head in disbelief and thought I was hearing things. As I continued to look out to sea I became aware of a stillness in me similar to one I experienced years ago while climbing towers for PG&E (in a very dissimilar context but the feeling was the same, see appendix D). Then the voice came again, this time clearer and I saw in my minds eye the whale that befriended me in Bodega, it was this creature that was speaking and said, "Urmas, it is time to return"!

That was it, no more, but I felt an immediate and all but forgotten kinship with the spirit of the whale and a delight in this mystical experience. When I mentioned it to Susan that night at dinner we both laughed at how strange it was and of course how I could not get engaged in whale work at that point in my life. I was one year away from fifty and besides, what possible avenue of entrance into the field was there for a man my age with no cetacean experience, no written papers and nothing particular to offer? Though my son was now attending The University of California, Susan had a son who was still a freshman in high school, so my time was hardly my own to spend chasing a fantasy.

A few nights later I had a dream in which I was a research scientist in a submersible looking out of a triangular shaped porthole. As I was gazing into the murky waters, I felt the presence of a monolithic creature nearby. Slowly it began to pass in front of the port (the only detail was that it was large and black) and then I saw it's eye. It was a whale, and there was no question but that it was looking at me. I knew in the dream that it was there to communicate something to me. What I heard in my dream mind was what I can only describe as a telepathic message, "it is time Urmas, there is work to be done". It was a very powerful dream and it's effect stayed with me the rest of the trip.

Upon returning to our home in Ukiah, California, I chanced to visit the library and talk with The Reference Librarian and good friend, Pattison "Pat" Hunt. He noticed my tan from my recent visit to Mexico and when I told him about my diving there he asked if I was still interested in marine biology (he was aware of my time at Bodega and my teaching at the college) and did "stuff" like that any more. I told him yes, I was still interested in stuff like that and why he had asked. He told me that he was on the board of directors of a fresh new NGO called Pelagikos, that was interested in engaging in marine work and that

I should meet their board president and Captain, Ethan Silva. He mentioned that they also had a research vessel that I might be interested in seeing that was berthed in Sausalito, a short two hour drive south. I said sure and Pattison offered to talk with The Captain and set up a meeting. Realize now that this was only a few days after coming back from my voice experience in Mexico.

The meeting was set up and I drove down to Sausalito to see the boat and talk with Captain Silva. As I walked down the wharf to the boat's berth I saw at the end of the pier a vessel that certainly stood out from the others. It was the 84' sailing schooner *Dariabar* and I realized it was in fact the boat I was to visit.

The 84' sailing schooner *Dariabar* that was eventually to become my home for six months. Here she is moored alongside pier four in Honolulu during a later expedition in 1998. I was to be christened as a sailor on "dog watch" while on the passage from San Francisco to The Hawaiian Islands that year.

Even for a non-sailor like myself it was obvious that this was a very special boat and had all the feel of a serious sea going research vessel. Over lunch in the salon we discussed each of our fantasies and it was soon clear that to a very real extent, our interests in regard to oceanic research were very similar. When I described my budding interest in whales and suggested Pelagikos engage in cetacean research, Ethan was immediately interested because that happened to be his passion also. When we parted company Ethan said he would talk to the board and see if there might be a place for me in the organization. Right at this time I had decided to go back into academia in pursuit of a PhD, but wanted to apply my experience of twenty years in the classroom towards a degree in Higher Education and Social Change rather than the sciences that had dominated my previous university experience. My substantial years in education and my street experiences socially and politically made this degree a natural fit for my interests at the time.

Ethan got back to me a week later and asked me to meet with The Board of Directors of Pelagikos and after a short introductory discussion I was somewhat surprised, but nevertheless delighted, to be offered the position of Executive Director of that organization. This was a non-pay position so I agreed to begin on a "time as available" basis and focus on generating interest in the marine sciences community in the use of our vessel *Dariabar* for on site research. Upon completion of the PhD I would devote more time to developing a research program of our own, and also developing a fund raising program to support our work. Both of these decisions, taking place during my fiftieth year of life, dramatically changed the direction of my journey and initiated my return to the whale... "the voice" had prevailed.

While I was being slowly and inexorably drawn into discovering Lilly's Cetacean Nation, the marine mammal research community was moving ahead with new methods and ideas regarding the scientific and social significance of the great whales.

I will mention some of these individuals and there work that I personally found interesting and that influenced my deepening connection with the great whales. There were, and are, many, many others who have worked devotedly in the past thirty or more years to elucidate more clearly just what these creatures represent. I am deeply indebted to their work and the inspiration I have found from it.

Roger Payne

Roger Payne can arguably be considered the "father" of cetacean studies, certainly in regard to vocalizations among the humpback whales. Payne began his studies on echo- location in bats but was drawn to other species and in 1967 made the discovery (along with Scott McVay) of whale song among the humpbacks. His work with the humpbacks led him into becoming a major spokesman for the campaign to end commercial whaling.

Payne described the whale songs as "exuberant, uninterrupted rivers of sound" with long repeated "themes", each song lasting up to thirty minutes and sung by an entire group of male humpbacks at once. The songs would be varied slightly between each breeding season, with a few new phrases added on and a few others dropped. He also suggested that fin and blue whales can communicate across entire oceans, which has now been proven for at least distances of over a thousand miles. Due to a great extent Payne's tireless work in this area of research was influential in the banning of commercial whaling in 1986. In 1971, Payne had founded the NGO "Ocean Alliance" which is active today in whale and ocean conservation. He has received many awards and provided endless hours of intelligent and entertaining insights with his media presentations. I will address the exciting and amazing new discoveries that have emerged from his early work in a later chapter.

Louis Herman

In 1970 Louis Herman, a psychologist/oceanographer at the University of Hawaii, founded The Kewalo Basin Marine Mammal Laboratory in Honolulu. His studies focused on bottlenose dolphin perception, cognition and communication. In 1975 he pioneered scientific study of the annual winter migration of humpback whales into Hawaiian Waters. His studies have generated over 120 scientific papers related to his work. In 1993 he founded The Dolphin Institute, a non- profit corporation dedicated to dolphins and whales through education, research and conservation. He also serves as a member of The Sanctuary Advisory Council for The Hawaiian Islands Humpback Whale National Marine Sanctuary. Dr. Herman added an entirely new dimension to Cetacean Studies and like Payne, his work influenced greater recognition of the truly remarkable character of both whales and dolphins. I was fortunate enough to meet Dr. Herman in 1998, while on expedition in The Hawaiian Islands with a group of six student interns. There is a wonderful story I will tell in the next chapter about a day with Dr. Herman and his dolphins and how something unexpected and delightful happened to one of my students that we were all gifted to witness.

Richard Sears

Richard Sears was born in Paris, from a French mother and an American father. His first encounter with whales was at 18 while on a training expedition aboard a schooner from Puerto Rico to Boston. He was, as so many of us have been, "entranced" by these dynamic creatures. In 1976, after completing a degree in biology he worked at Woods Hole Oceanographic Institute where he was privileged to work with the blue whales of Baie de Moisie. Later, Sears became a naturalist on board whale-watching vessels in Massachusetts and worked alongside some of the pioneers of whale research including Roger Payne. From them, he learned that in order to know whales one must spend a lot of time at sea with these giants. In 1979 Sears returned to Mingan to exclusively study the blue whales of that area. There he established The Mingan Island Cetacean Society, Inc. to provide new incentives on cetacean studies. His work over the past twenty plus years with the blue whales has been exemplary of the devoted and passionate scientist. I met Richard in 2004 while on a personnel journey of discovery after the end of my third marriage, the death of my mother and the specter of cancer. I feel privileged to have met him and I owe him a great debt for inspiring me to seriously return to the study of the whales in my area the following year.

John Calambokidis

John Calambokidis was also one of the first research biologists to begin serious study of whale behavior. In 1979 he founded Cascadia Research, a non-profit organization in Olympia, Washington and has, since that time, directed over 100 projects, written two books and produced over 150 papers regarding marine mammals. He is also periodically on the adjunct faculty of Evergreen State College in Washington. John's work focuses on the biology of marine mammals and has contributed a wealth of information regarding the human impact upon them. His book, *"Blue Whales"* with G.H. Steiger is a classic in blue whale literature. Most of John's work takes place in The North Pacific where he has extensively studied the blue, humpback and gray whales of the region between Alaska and Central America. In 2009 his work was featured in a National Geographic article and television special. I never met John but I was doing ID work in The Channel Islands one year where our boat wakes passed as each of us were on a search.

Chris Clark

Chris Clark of Cornell University's bioacoustics lab has done as much or more to bring the subsurface behavior of cetaceans literally to light than anyone engaged in whale research. In so doing he has also demonstrated via acoustical studies, the intelligence and complex social patterns exhibited by these creatures. His particular focus has been the sounds of the blue, humpback and minke whales of The North Atlantic. Using the declassified antisubmarine Sound Surveillance System (SOSUS) made up of a non-intrusive hydrophone array he is able to "listen" in on the myriad of whale sound and relate it to individual whales, groups of whales and distant singers. One of his major concerns has been the possible negative effect of ambient ocean noise (what he refers to as "Ocean Smog") on the abilities of the great whales to communicate effectively. Unfortunately this noise is doubling every ten years and is now considered a major problem in whale conservation. Clark is The I.P. Johnson Director of Cornell's Bioacoustics Research Program. In 1998, Dr. Clark and his team used our (Pelagikos) schooner *Dariabar*, as their research platform during The ATOC experiment. It was during the five months we spent on station in Hawaii that I shared time with him while crewing on board our vessel.

Jorge Urban Ramirez

Dr. Jorge Urban Ramirez was born in Mexico City in 1956. He received his PhD from the National University of México (UNAM). He is a research specialist in marine mammals of México. Jorge has presented papers on this topic at more than 60 international meetings, and has written more than 40 scientific publications about the great whales and dolphins of the Gulf of California and Mexican Pacific. From 1991 to 1993 he was president of the Sociedad Mexicana para el Estudio de los Mamíferos Marinos (SOMEMMA). He is a member of the Scientific Committee of the International Whaling Commission and the Cetacean Specialist Group of the IUCN. Jorge is the Coordinator of the Marine Mammal Research Program of the Universidad Autónoma de Baja California Sur (UABCS) in La Paz, México. For the past eight years I have been privileged and honored, to be designated as a "Collaborador" in Dr. Urban's studies here in Mexico. I was invited to join his team in 2005 after I sent him my humpback photo ID's of that season that I had taken on The East Cape of The Sea of Cortez. His work involving the migration patterns and the population dynamics (particularly breeding and calving activity) of the humpback whales in Mexican waters are extensive, and have resulted in a proposal to soon develop a Marine Mammal Protected Area (MMPA) here in Baja California Sur.

Each of these men became, in a sense a mentor of mine, especially Richard Sears and Jorge Urban. These men encouraged me to re-engage in whale research, thereby inspiring me to actually live out my fantasy. To them I owe more than can be adequately expressed. In a later chapter I will describe some of the discoveries, insights and theories that all of these men have contributed to what I refer to as "Whale Lore".

Chapter IV
Ireland, Shamans and Whales

In the previous chapter I described my sudden elevation to the position of Executive Director of Pelagikos, and the responsibility for developing a research program focused on the great whales of The East Pacific. To a great extent too, I needed at that time to engage in some serious fund raising. In retrospect I now see that I was not really fit for the job because I was not experienced at all in whale studies and I was a terrible salesman. By that I mean that I was never very good at selling people on investing their money. Even as a Cub Scout I was embarrassingly timid about asking people to buy raffle tickets each year. The same applied in Boy Scouts and any other activity that I engaged in that required a request for money. In other ways I look back at that period with some satisfaction in that I did design, with the creative help of my good friend David Smith of Ninetrees Design, an attractive brochure to advertise our offering to the scientific community. I also organized and directed three student expeditions to The Channel Islands, off the southern California coast, to investigate the island's ecosystems and to do some blue whale and humpback ID work there. In 1998 I took six students to Hawaii for a ten-day sail around The Hawaiian Islands to study the local ecology there.

It was during these expeditions that I had the opportunity to enrich my Cousteau fantasy but more importantly in the long run, re-establish personal contact with the great whales. Again, at this time I was not remotely aware of the deep connection I would make with these wonderful creatures. It was a rather odd event that gave me the chutzpah to go ahead with this new direction my life seemed to be taking. After all, it was not so much a desire to actually discover scientific facts about the whales that drew my interest, it was hearing "The Voice" and realizing through the little research (primarily reading the studies of the men mentioned in the previous chapter) that I was keeping up on, that these creatures may truly be sentient on a level yet to be fully recognized and appreciated. This aspect of the whales did interest me but yet again, not particularly strongly.

What did, in a very real sense, focus my interest in whales, and indeed gave it a new and more profound influence in my life, was an Ecology Conference in Ireland. In early 1992 a brochure crossed my desk describing an International Transpersonal Association (ITA) conference that was to take place later that year in Killarney, Ireland. As I perused the brochure I recognized many of the names of the speakers and realized that they were the very people who had been shaping my personal philosophy regarding the environment and the difficulties we would be facing as a species in the near future. They were also the people who had been instrumental in encouraging me to look deeper into the spiritual aspects of life on earth. At this point I had been teaching environmental science for almost twenty years and had developed a deep concern regarding what was clearly an over-exploitation of the planet's natural resources and that could only result in serious environmental consequences in the future. The conference was to address these issues from a wide range of disciplines, including the spiritual, and I was intent on attending.

The problem was that I did not have the money to fly to Ireland, rent a room and eat for the ten days of the conference. This time, however, I was able to ask for money and somehow, with the help of a great many friends, family and students, I was able to obtain just enough funds to allow for the trip. I was also lucky in that the dates for the conference happened to fall on an academic vacation so I had the free time to go.

Waiting for the plane at San Francisco Airport I again went over the individual speakers and the titles for their presentations. Here are a few of those:

> **Awakening the Heart of Buddha in the West**, Jack Kornfield
> **The Western Mind At The Threshold**, Richard Tarnas
> **Riding the Wave of Change**, Ram Dass
> **Introduction to the Spiritual Path for the Scientist**, Charles Tart
> **Creation Spirituality: A Movement of Hope**, Matthew Fox
> **Science, Spirituality, and the Present World Crisis**, David Bohm
> **Change, Conflict, and Resolution from A Cross Cultural Perspective**, Angeles Arrien
> **Synchronicity and the Tao**, Jean Shinoda Bolen

Morphic Resonance and Collective Memory,
Rupert Sheldrake
Food of the Gods: Searching for the Original Tree of Knowledge, Terrence McKenna
Science and Spirituality: The Gap Closes,
Robert Schwartz

These and other papers given by Winona LaDuke, Elisabet Shatouris, Barry Commoner and Thomas Pinkson were to fill my days in Ireland with intellectual and spiritual discoveries and awakenings beyond my wildest dreams. All the while also reflecting on what this new awareness meant for the journey I was about to partake in with Pelagikos and the great whales. I made a point of personally meeting a number of the speakers and discussing with them the work that I intended to begin with the whales and how their presentations had moved me to a greater understanding of just what it was in fact that I was going to do with this opportunity I had with Pelagikos. Four people were particularly interested in the possibilities I had before me, and two of those eventually accompanied me on expeditions to The Channel Islands off the Southern California coast tracking blue whales.

It was definitely the deep connections I made with these four people that dramatically influenced the tenor of my personal interest in the great whales and thereby was also reflected in how we at Pelagikos related to the whales while in the field and in their presence. I will tell two stories here regarding Ram Dass and Thomas Pinkson, both of whom had a most significant influence on my vision for what I would do.

One evening in Killarney I was walking home from a hole in the wall fish and chips place (extraordinary by the way) when I heard some Irish music playing in a little outdoor park. I sat and listened to the music for some time and saw another man there that was clearly not Irish and I was certain was attending the conference too (he looked...Californian). We happened to leave about the same time so I introduced myself and we discussed my work as we walked back to the hotel area. As it turned out it was Thomas "Tomas" Pinkson, one of the presenters at the conference. Thomas had his PhD in psychology and more importantly to me, had successfully completed a twenty-year initiation in The Shamanic Knowledge of The Huichol Indians of Mexico. I was fascinated with his experiences and on impulse a year later invited him on one of our first expeditions to The Channel Islands.

On another day at the conference, I was going into a room where Barry Commoner, the environmental activist, scientist and 1980 American Presidential candidate, was going to speak. I was early and the room was almost empty but for one man...and that was Ram Dass, a man whose book, *"Be Here Now"* in a sense changed my life. I could not let this opportunity go by and though I knew it was rude, I approached him, sat down next to him and explained to him my high regard for his mind and soul and how important his words had been at a time in my life where very little was making real sense. He made me feel immediately at ease and we began to discuss the great journey we humans were involved in and how there was so much to be done. I told him about my interest in the whales and how I was beginning to feel that there was so much more to know about them in a spiritual sense.

As the room began to fill I knew I had to say good by and let him be. When I stood up to go, he also stood, put his arms around me and with tears in his eyes said to me, "you are doing The Good Work", we hugged with great gusto and I left.

There are many stories to tell regarding my experiences at that conference. Indeed, I think I could write an entire book about those ten days alone.

Understandably, upon my return to The States, the college and Pelagikos, the energy from the conference rapidly propelled me into a clear recognition of what I wanted to accomplish through Pelagikos; I wanted to see (metaphorically) into the "Eye of The Whale". How that was going to happen I had no idea. I just knew that for me personally this was going to be a spiritual journey as much as a scientific one.

After completing the PhD in early 1993 I was able to devote more time to my position with Pelagikos and I found myself occasionally at sea on our schooner *Dariabar*. I had no experience as a sailor but I went along on a number of short expeditions that now put me closer to the whales than I had ever been. Some interest had been generated within the scientific and military communities so Pelagikos began to receive small contracts to act as a research vessel for their studies. Most of these were bio-acoustic studies and some photo identification work. I sailed on her to The Farallon Islands off of San Francisco and saw, up close (a few hundred feet) for the first time, both humpback and blue whales. Although I was duly impressed with their size and spirit, I was also a bit seasick and did not fully appreciate the time with them. This (seasickness) was to become a problem for me throughout my time on *Dariabar* and unfortunately prevented me from enjoying my years with Pelagikos as much as I could have. Nevertheless, I went through that entire time without The Captain or the crew ever knowing of my discomfort. Being the grandson of a famous sea captain made it imperative to my pride that I not lose it, literally as well as figuratively! We also did some work in Monterey Bay and along the northern California coast. One contract did take *Dariabar* to Mexican waters where she was a laboratory platform for a number of different agencies and I first came across the name of Dr. Jorge Urban. I did not accompany the boat nor meet Dr. Urban at that time due to my other responsibilities.

While *Daraibar* was away in Mexico I took a road trip through The American South West and visited a man who had e-mailed me in regard to Pelagikos and whale studies. His name was Scott Taylor and we immediately became friends for we had the same dream...looking into the "eye of the whale". We talked for many entertaining and inspirational hours about our dream and how we could best implement it. What was clear was that both of us were more spiritually attracted to the whales than scientifically. As a devotee and friend of the famous John Lilly, Scott had been contemplating the dream for some time. Eventually he moved to Australia, started an institute to study the dolphins (I eventually became an advisory board member) and then wrote his excellent book, *"Souls in the Sea"*.

I returned to California after this trip with an increased passion to know these grand creatures. He also introduced me to the term, Cetacean Nation, that I began to use occasionally when referring to whales and dolphins. This term had originated with John Lilly and was meant to imply a higher order of intelligence to the whales and dolphins and to accept them as a "Nation" that should be recognized and respected...not hunted and killed.

I thought then that a natural and valuable use of *Dariabar* was to use her as an educational platform for students interested in Marine Sciences and particularly in regard to whales and dolphins. During the spring of 1995 and 1996 I organized and personally conducted three student expeditions to The Channel Islands to investigate the humpback and blue whale populations there and to study the general ecology of the area.

During the months of July, August and September, krill forms dense swarms near the shelf break off The Channel Islands, where the mixing of the warm waters from the south and cold waters from the north promote the growth of the plankton they live on. Krill (small shrimp like creatures) are a favorite food of all baleen whales and are known to be the main attraction for migrating blue and humpback whales in The East Pacific.

With their (blue whales) worldwide population reduced from around 300,000 to now under 10,000, they remain endangered. Despite this decline, the Santa Barbara Channel is probably the best location in the world to see blue whales. Up to several hundred of the 2,000 - 3,000 that feed off the California coast can be found feeding in the channel during the summer.

I will relate two stories from those expeditions that dramatically changed my personal relationship to the whales and set the tone for my later encounters in Mexico.

In 1995 I contacted "Tomas", my friend from The ITA Conference, and asked him to join us for an expedition to The Channel Islands. Though he was somewhat hesitant due to a similar predilection for seasickness as myself, he agreed to join us and act as a "Spiritual Advisor" while we were in the company of the whales there. After sailing across the channel from Santa Barbara the first day, we anchored off of Santa Cruz Island where we gathered at the stern cockpit for a welcoming ceremony for the whales. Tomas burned some sage and began drumming and chanting to connect with what he called "The Whale People". Then he asked each of us to express why we were there and what we were willing to give to the whales. This had never occurred to me before...**what could I give to the whales**?

It is difficult to even imagine being able to give something to the whales. It's obvious immediately to anyone, that it would not be something material. Perhaps it could be something physical nonetheless, as in helping establish some form of protection, i.e. conservation, of their species. The fact that they were endangered was no secret at this time. Could I do something to return the life saving favor that the humpback did for me back in 1971 in Bodega Bay? Tall order indeed, yet perhaps, just perhaps, my work within Pelagikos could eventually result in some kind of tangible benefit to their life ways. One thing that Tomas made clear in his advice to us for the next days search was the necessity for us to tune our spirits to the common thread that existed between ourselves and the "The Whale People" (see appendix D for another story regarding Tomas and myself). We all made a commitment to do that and the next day dawned clear and bright in the channel, and as we sailed out from our anchorage off Santa Cruz Island and began NW in the direction Tomas suggested we go our expectations were high and our spirits clear and friendly.

As we searched across the unusually calm waters of the channel that morning, we suddenly sighted the back of a blue whale approaching directly to us from about a half-mile to starboard. We decided to drop sail and simply drift as the whale continued on this apparently direct line toward us. The closer it came the more excited we became and as we cheered it on, Tomas sat at the stern in both childlike wonder (he had never seen a whale before) and deep Shamanic reverence. The whale slowed and with inspiring mass and grace, slid past the stern just inches away from the rudder of *Dariabar*.

We were all awed by the experience, not just it's size, which was substantial...65' plus, but it's calm and graceful swim so close to our boat, and nary a ripple! During the rest of the day we saw no more whales but in our evening gathering with Tomas he told us what he felt had transpired and what we might expect the next few days. Here is his story as expressed in a letter/article he wrote to me afterward.

Sacred Ecology, Whales, Mosquitoes and a Flowing Crystal River, Tom Pinkson, Ph.D., Sacred Ecology - from Killarney pubs to California's ocean

It started outside a rollicking pub in, Ireland. It was about one in the morning and I stopped to listen to some wonderful Irish music as I made my way back to my hotel from a presentation I'd made at the International Transpersonal Psychology Conference. While the official proceedings had ended for the night, the unofficial ones were just beginning to open up. Several fellow early revelers stood on the sidewalk enjoying the scene and talking with one another. I connected with one man in particular, a fellow Californian, and we exchanged business cards, then went on our way.

Six months later I was surprised to hear from him when he called me on the phone and asked me if I would be a guest teacher for <u>Pelagikos</u>, a program he directed that involved tracking blue whales using their eighty-four foot sailboat equipped for research purposes. I would spend one week with the crew and young students sailing off the Channel Islands near Santa Barbara. My job was to "bring spirituality into the scientific enterprise" and to use shamanic ways to help connect with the spirit of the whales. Excitedly I said "yes", and then turned to face my propensity for seasickness.

The first night out we anchored off one of the Channel Islands. That night the crew and students gathered on deck under ample moonlight for our first ceremonial work. I asked the members of our party to think about a gift they would like to bring to the whales. "Most people take when they come to sea with little thought of giving something back. I get that our task here is to give something back, even if we don't see any whales." I emphasized.

Then I picked up my drum, purified it with the smoke of burning sage and offered the cleansing smoke around the circle to the others. While it was going around, I began a soft beat on the drum and began to pray. "Thank you Great Mother Ocean. Thank you for the gifts of your wisdom being. I pray for your health and the health of all the swimming people. Help us to open to your Medicine teachings, and help us to see what it is we can do to protect and honor you".

The following morning I did another drum journey, this time carrying the spirit of the gifts down into the sea. The beat of the drum carried me to a cave in the depths of the ocean. Inside the cave was a counsel of Whale Elders. They were in a circle surrounding a fire and seemed to be expecting me. I introduced myself and offered a piece of wood to feed Grandpa Fire. One Elder purified me with sacred smoke and motioned me to speak. I told them I had come bringing thankfulness prayers for their being as well as prayers for their health and protection, especially for their children and old people. Also that I came with gifts from others. One by one I introduced each member of our party who came up and shared their gift.

When all the gifts had been shared, I asked the Elders to consider sending us some messengers on the next day when we went out to a channel between the islands where the biologist on our crew thought there was a good chance we would find whales. "I do not know if it is for our greatest good, or for yours'," I said, "that we come into contact with you tomorrow. If it is not, I accept that, and will still continue to pray for your lives and give thanks for your wisdom teachings, for you are some of the oldest people on the planet and you know the history going way back to the earliest times. We open to whatever your counsel feels is best for us all".

The next day we sailed off with hopeful anticipation. The biologist sat high up in an observation chair with her trusty binoculars. The boat, Dariabar, sailed smoothly through the sea. We were in open waters now, the only land was far away. Suddenly an excited cry came from above; "Spouts three hundred yards off the starboard bow!" "They're humpbacks" she yelled with glee. The skipper turned the boat in the direction of the spouts and off we went. Through my own binoculars I could make out the contours of these huge masses undulating through the waves, then disappearing into the depths only to appear again in five or six minutes somewhere else.

I was thrilled. I'd seen gray whales migrating up and down the California coastline before but that was always when I had been on shore watching them from a land base. Now I was in their home, in their terrain, and it was an entirely different experience. The ocean was alive and breathing, constantly changing in mood and cycles of wind, wave, color and sound. The whales knew what was happening and danced with it in graceful T'ai chi like movements that were millions of years old. It was a bit frustrating though, because each time we approached them, they would disappear into the depths and reappear further and further away.

Two Blue Whale People Surface

Then it happened. Something the biologist who had been tracking whales for years said was not unheard of but was very "unusual". Something that stunned us all, then sent shivers running up and down our spines amidst bursts of gleeful shouts of excitement, joy and amazement. Two blue whales, longer than our eighty-four foot boat, surfaced just ten feet away and began to slowly circle the boat! Speechless in awe, I watched these magnificent beings who are the largest animal that we know of to have ever lived on Mother Earth, or Mother Ocean I should say, do laps around our vessel checking us out. "Unbelievable" was all I could mumble as I and the others ran around the boat following the whales, taking pictures, taking in the immensity of their being and that they are intelligent creatures who I felt had been sent by the Elders Counsel from the drum journey.

I sent them my love and thanked them for coming to us. I opened myself to their means of communication and let my body-being take in what ever they had to send. I felt communion with them and that they were showing me a more graceful way of being than my culturally conditioned push-force-work hard to make things happen mode of being. I felt the ancientness of their way and that they were messengers:

You have to learn a new way of living, you two-leggeds. You have to relearn how to respect the Sea of Being which is the ocean we all swim in. We are all connected. We are all related. We are all equally loved by the Creator of this Great Universe. You have forgotten and live your lives as if you were the most important of God's creations. But as you can see here, you are not. We must all learn to work cooperatively, together, for the good of all or soon great changes will come that will reduce your arrogance back into a respectful humility which is required for all of us to live together in harmony and happiness. Remember us, your swimming relatives, and help us to care for our home which is really your home as well.

Then they left. They swam off away from the boat and while we saw them and enjoyed sailing with them throughout the rest of our time at sea, never again did they come that close. No, their message was delivered, and then they went on their way. As I write this now, I'm back on land, but I still feel connected to the Counsel Fire of Whale Spirit Elders and their special gift to us. I feel a telepathic line of communication has been opened and their presence is as close as my breath.

Tomas

When the expedition was over there was a palpable change in all of us and I knew it would live within us for years to come...indeed a lifetime.

The second story I would like to tell has to do with another expedition the following year in 1996, this time with our guest scholar Elisabet Satouris; biologist, author *of "From Chaos to Cosmos"* and a speaker at The ITA Conference I attended and befriended in 1992. Interestingly enough, she joined us shortly after she had completed a workshop in The Peruvian Andes with an Inca Shaman, and when she had asked the shaman how she (we) should approach the whales he told her we should chant, sing, dance and be full of joyous energy in their presence. On the first night aboard, as we were docked in Santa Barbara Harbor, we had a now (due to Tomas's encouragement) traditional cockpit gathering and each asked, "what do I want personally to accomplish on this trip". I do not remember the wishes of the others but I remember mine vividly, I wanted to see into The Eye of The Whale, again a metaphor for seeing into the depths of what I now too began to think of as "The Whale People". Elisabet shared her message from The Inca Shaman and we all agreed that there was no reason at all that we could not combine our scientific work (photo Identification) with a joyous celebration of respect and commonality.

The first day's sail across the Channel was an exciting one, due to high winds and relatively rough seas, but *Dariabar* could easily handle the conditions and we arrived at San Miguel Island in good order. We anchored on the lee of the island and prepared for the next days sail south east to the area we had begun to call "Whale City" due to the number of both blue and humpback whales we had encountered there before.

The following day conditions could not have been better for spotting whales and we were all in a fine mood. Elisabet had brought her Australian didgeridoo and we had a number of drums on board to create the festive mood Elisabet's Shaman had suggested. We also had a hydrophone rigged to *Dariabar's* stereo unit and speakers set in the lab so we could detect any whale sound nearby. There was very little wind so we motor sailed from San Miguel Island towards the area between Santa Rosa and Santa Cruz Islands that was our "Whale City". As we entered the area we became more and more vigilant and excited. Within just a short time we saw two blows behind us and from their shape and height we could tell that they were from two humpback whales about a quarter mile behind us. The excitement increased as it became clear that these two whales were actually following us and were clearly intent on closing the gap between us. We shut off the engine and drifted with the light wind as the whales came closer and closer. Finally they settled behind the stern so close that we could almost lean over and touch them. This is relatively common behavior among gray whales but not humpbacks. As we all lay on the deck and got as close as we could to these two whales they kept right along with us and stayed immediately below as we humans hung over the edge. At one point they rolled on their sides and looked at us and for the first time in my life I looked into the eye of the whale from close up and realized I was looking into the eye of another intelligent species. I felt that there was complete and utter recognition in their eyes of the commonality between us, and I am certain they saw this in our eyes too. Suddenly Captain Silva's young son turned to me and said, "Urmas, you are looking into the eye of the whale". I am not sure what moved between those whales and myself at that moment but I do know that I had been touched by the thoughts and feelings of another of earth's creatures that had a heart and soul and knew of their own existence. It would be some time yet before that experience and all it's inspiration would become the focus of my later life's work, but it was a focal moment.

Later that same day we heard a number of whales vocalizing in the distance. We could also see a fair number of blows from both humpbacks and blue whales within a mile or two of our position. One of the students on board began drumming and Elisabet placed her diggeradoo next to the hull (*Dariabar* has an all steel hull) and began playing it, mimicing the sounds we were hearing from the whales. We were quite certain that the dominate sounds were from blue whales and we could see a number of them moving parallel to us about half a mile away. I was on deck with binoculars scanning the distance when I realized that the blue whales were changing direction and coming our way. They had made a 90 degree turn and were unquestionably coming directly to us. I called to the others to come see this and soon all the students and crew were on deck and entranced by the sight of these enormous creatures collecting around us. Soon there were at least seven blue whales surrounding *Dariabar* and swimming around the boat. It became evident that the whales were forming a ring around *Dariabar* and tightening it so that they were nose (rostrum) to tail (fluke) as they circled us. Everyone was so excited that our camera man ended up filming the deck, Elisabet had tears in her eyes, The Captain smiled and remarked, "they are playing with us", our helmsman literally ripped off the amulet he had around his neck and flung it into the sea next to the whales and then closed his eyes as his hand on the wheel moved in harmony with the wakes of the whales, and I was literally transfixed as I felt the energy of these playful, sentient creatures slowly begin turning *Dariabar* as if we were in a vortex. It was all that Elisabet's shaman had seen...it was a party!!! A party, I believed, to celebrate the commonality between human and whale. They stayed with us for only about a half hour but in that time all of us felt a change in our personal realities. Sadly our on board photographer was so taken by the moment that she ended up only filming the deck of the boat during the entire episode.

Those expeditions to The Channel Islands were instrumental in introducing me to what John Lilly called "The Cetacean Nation" and the depth of intelligence found within these creatures, but there was a long journey yet for me before I would become intimately connected and that would inspire my present work and the theory I have developed about them. There was much The Whale People had to teach me before I was brought here to Baja California Sur to complete my time on this Earth in their august company.

Chapter V

Life Changes and The ATOC Experiment

During the fall of 1997 I went through some difficult relationship problems with my third wife and decided to move out of our home. I found a small rental in the country and struggled with what I would do next. In order partly to lessen the disappointment and sense of yet another marital failure I purchased a sit-on-top kayak and began spending more time in the waters of Mendocino Counties lakes, rivers and ocean. It became a wonderful salve to my damaged pride, ego and heart.

Then in October of that year Captain Silva informed me that he had engineered a contract with Cornell University for *Dariabar* to be their lab platform in The Hawaiian Islands during The ATOC Experiment. This was an experiment devised by The Scripps Institute of Oceanography for determining ocean temperature within the North Pacific Basin to see if indeed atmospheric temperature rise was reflected in ocean temperature rise. The world's oceans act as a gigantic sink for excess atmospheric temperature and carbon dioxide and there was justifiable concern that ocean temperatures were rising inordinately fast and that this was a reflection of the then new (to the general public, but contemplated and predicted by the scientific community as far back as 1956) specter of "Global Warming". There was no better way to determine whether the atmosphere was indeed warming than to check if the oceans were warming in their attempt to buffer the atmospheric rise already seen. A clever way to do that in this very large body of water was to set off a loud sound underwater and then have a number of monitoring stations throughout the basin record the time it took the sound to reach them. This would give Scripps a clear picture of the speed of the sound in the water, and since that was determined by density of the water, and the density in turn, was determined by the temperature, a definitive picture could then be seen of the actual temperature of the entire ocean basin. A similar technique is used by geophysicists to determine the boundaries between different layers of the earth's interior. There was a problem however that was immediately recognized by Chris Clark of Cornell's bioacoustics lab. The area where the big sound would be emitted was the very same area that was used by 50% of the breeding and calving humpback whale population of The North Pacific. Dr. Clark's concern was that "The Big Sound" would negatively effect the normal patterns of the whales and perhaps even damage their very sensitive hearing, or in the worst case, seriously effect their health and perhaps even their lives. Therefore, built into the project, was a six month on-sight monitoring of humpback behavior during the ATOC Experiment, and the responsibility for that monitoring was given to Dr. Clark and his team from Cornell. We, Pelagikos, because of the sea going versatility of our sailing schooner *Dariabar*, were given the contract with Cornell to be their lab platform on-sight in Hawaii as they carried out the monitoring.

Captain Silva asked if I would like to accompany *Dariabar* for the six-month sail as a crewmember. This would mean a long ocean "passage" and for me a serious confrontation with my worst fears regarding the ocean, and of course, the problem with seasickness. Nonetheless, I realized that I could join the expedition in San Francisco immediately after the fall semester of teaching, miss spring semester, but make up for it by teaching summer school. The timing couldn't be better, especially since I really needed something to get me out of my depression regarding my failing marriage.

Impulsively, I agreed to go. I packed up a very few essentials (my berth was in a small cabin on the port side of the vessel measuring 7' by 5', comfortable enough, but as tight a quarters as I had ever experienced) and joined the crew on *Dariabar* in November of 1997. I was allowed to take my kayak, now named Haldjas (after The Estonian Goddess that protects all beings in nature) on board secured to the stanchions on the starboard side of *Dariabar*. Due to other responsibilities of The Captain, Christmas and a delayed arrival of a crewman coming from New Zealand to join us, we got a late start and finally departed the harbor in Sausalito on the 28th of December, sailing right into an El Nino.

Normally at this time of year The Trade Winds blow off the California coast and it is a literal sleigh ride to Hawaii. In an El Nino The Trade Winds reverse direction and blow against the coast thereby making it necessary to either sail directly into the wind (impractical if not impossible) or sail along the coast as far south as is necessary and catch the North Equatorial Current below the winds of The El Nino and then sail west to Hawaii. To accomplish this we sailed south almost to Cabo San Lucas, Mexico, before we could safely cross below the contrary winds that were now blowing strongly in The East Pacific.

We had missed the annual whale migration south due to our late departure, and when we did start sailing west it became abundantly clear that the humpbacks of The East Pacific did not cross from the mainland to Hawaii as part of their yearly migration (although it is now known that a very few do), we saw only one whale during the entire passage and that was a single sperm whale. We did collect some plankton samples for a colleague of mine during the crossing but otherwise concentrated on arriving in Hawaii on time to meet Dr. Clark and his team on the island of Kauai where we would headquarter during The ATOC project.

The total passage took us two weeks of mostly tedious motor sailing due to the unpredictable status of the winds created by The El Nino. On the tenth night out and pretty much in the exact middle of The East Pacific we were hit by a midnight squall with winds gusting at 55-60 mph...a gale. As the most novice sailor on board I had been given "Dog Watch" (12:00 midnight to 4:00 am and 12:00 noon to 4:00 pm) by The Captain, and along with my helms mate Jack Frost, a young ex New Zealand Commando, we were responsible for manning the helm during the storm. When Jack and I came on watch we still had 4,000 square feet of sail up (trying to catch as much wind as possible during the minor winds of the day) and as we hit the deck for our watch, the first mate, Charlie Patton, suggested we be prepared for a blow. The blow was more than expected and within minutes of our taking over the watch we were in the storm. For a period of about a half hour all crew were on deck in a pitching sea (20' swells cresting all around) trying desperately to get the canvas down as *Dariabar* heeled over at 45-50 degrees almost dropping our mainsail boom into the pacific. As they used to say, it was "all elbows and assholes" for that half hour but we got it done. I was at the helm and though I tried I did a poor job of sailing *Dariabar* while everyone else was helping drop the sheets. I don't recall ever being so scared in my life. I couldn't see because of the stinging rain in my eyes, my glasses had to be put away because I couldn't of course see through them, my hat had flown off and my hair was in my eyes, it was pitch dark, the sea's around me were like a living breathing monster, the wind shrieked through the rigging and there were sails flapping everywhere. Eventually the first mate took over at the wheel (with a great deal of cursing at my ineptness, which came to be one of the most hurtful experiences in my life) and I went forward to help lower the remaining sails. We hadn't had time to harness in so we were all working on a wet, heaving deck during the whole operation without safety lines. I took position on the starboard side next to the main mast and leaning into the sea with my butt and back against two 3/8" cables, waited for the queen sail to come down so I could secure it to the stanchions on my side of the vessel.

It was at this moment that I went through a classic time warp. To me, the wind suddenly could not be heard, the sea became a soft undulating non-menacing bed and I felt relieved of all concern...I was at peace-I was without fear! Then the queen sail came down and collapsed over the top of me and The Captain was shouting to find out if I was alright or had been knocked overboard by the force of the falling sail. I woke immediately out of my dream state and yelled back that I was OK and commenced to help tie the sail down.

After all the sails but the jib (needed to keep our bow to the wind) were down and secured, the rest of the crew went below while Jack and I still had the helm for another three hours or more. My fear returned when the crew went back below and putting all pride aside, I asked that Jack stay with me by the wheel on the storm bridge rather than take comfort in the chart room where it was his right to go while I had the wheel. Much to my relief, and what inspired a very close friendship during the rest of the project, he responded with a simple "thumbs up" and we rode out the storm together for the next 3 hours. When it was his turn at the wheel (2 hrs on, 2 off) he suggested I go below and relax but after his courteous behavior I chose to also stay by him (though he didn't need me) for his 2 hours and we finished our watch as "mates".

What I realized from that night was that my fear of the ocean stemmed back to the escape across The Baltic Sea when we too hit a storm and as a three year old I was terrified. That fear stuck with me but this time as an adult I could face the fear, conquer it and move on. That night, in that storm, I forgave the little boy that was in me, and the next morning I was literally a new man - my fear of the sea was gone.

Urmas Kaldveer, PhD

The ATOC (Atmospheric Thermoptery of Ocean Climate) project kept us in Hawaiian waters for the next four months as we took station each morning over the Big Sound Box (eight miles out from Kauai and 8,000 depth) at 7:00 am and followed any group of humpbacks that were in the immediate area to monitor their surface behavior and vocalizations. The experiment was designed so that we never knew exactly when a sound blast would be generated in order that our data was not skewed towards behavior we thought was anomalous. We kept on or near station until 4:00 pm at which time no more sounds would be generated.

Later all our observations would be examined by a team comparing those observations with the timing of the actual blasts and an evaluation done regarding any effects upon the whales. We also had on board, a twenty seven hydrophone array, recently declassified as a listening device for enemy submarines. With this very high tech array and our on board computer monitors we could observe the position and vocalizations of individual whales and their groupings. Any anomalous behavior could be seen in real time by watching these monitors. We also set out a group of a dozen or more sona buoys to listen 24/7 on any humpback vocalizations while the big sound was turned off. This was done in order to establish a baseline norm for vocalizations in the area. These buoys consisted of a plastic sphere containing a high tech computer hard drive and hydrophone set up connected to a heavy anchor and dropped down to the bottom of the sea. These would be "popped up" later by using a transponder to send a burn message to the connection between the sona buoy and it's anchor (the anchors were expendable, they were five gallon cans filled with concrete). Over the period of the experiment we only lost three buoys, and though each one was a $7,000.00 investment, Dr. Clark felt that our (Pelagikos) efforts had been commendable.

During the ATOC experiment we dropped these sona buoys off the coast of Kauai to record whale sound. Inside the yellow sphere was a very sophisticated computer and microphone to hear and record their vocalizations for later analysis.

We would stay on station for ten days then return to Nawiliwili Harbor on Kauai to re-supply and take on new groups of student interns working for Dr. Clark and his onsite director, Dr. Adam Frankel. It was during those four months of sailing that I became more involved in the scientific side of my personal journey with the whales, yet two experiences I did have were clearly of a more singular and personal nature. I might add that our results from those days on station showed that the whales in the area had not in fact been negatively effected by the experiment and that indeed they seemed to have been more curious than annoyed. Ironically what we did see was that what really upset them was jet skiis! (we often followed a group of whales closer to shore where tourists played). This was to play a role in my later push for restrictions on these vehicles wherever whales were found. I will talk more on that later.

One of the more personal experiences that I referred to took place shortly after the passage while we were still on The Big Island of Hawaii awaiting the hydrophone array. One day the weather was ideal for kayaking so I asked Danny, our radio operator, to pull my kayak behind the Zodiac (a 14' motorized dingy) out of the harbor and into Hilo Bay so I could attempt a whale encounter. He left me about three miles out and I began paddling randomly in the chance that I might meet up with a humpback. Before too long I did see a humpback moving parallel with me and about a half mile off. I started an intercept route and as we started to come closer together I realized that this was the closest I had been to a whale while in my kayak. That is considerably different than being next to a whale in an 84' sailboat as I had been numerous times in The Channel Islands. As I pulled closer to it I began to hear it's blows and also realize just how big this creature was. I must admit I hesitated slightly but then with my new relationship to the sea, I felt the will to keep getting closer. When I was within 2 or 3 hundred feet of it a sport fishing boat went by between us, the whale sounded and I was alone. Though I have had dozens of "close(er) encounters" in my kayak since then, and far more dramatic, it was my first.... and you know what they say about that.

The other personal experience took place at the end of the expedition when we had six students join us from California for a ten day sail/seminar in Hawaiian Island Ecology

There were a number of delightful moments with these students but one stands out that lent a special flavor to what had been a purely scientific quest. By necessity throughout the ATOC experiment, we (Pelagikos) were responsible for supporting the scientific efforts of the Cornell University team. There was very little time or opportunity for me to engage in a personal quest for seeing into the "eye of the whale". This was true during the ten day seminar also, but a chance meeting with Dr. Louis Herman at The Kewalo Basin Marine Laboratory on Oahu changed all that and provided me and my students with a unique opportunity that in it's own way allowed us to see into the eye of the whale.

I had contacted Dr. Herman weeks before and explained our (Pelagikos) role in ATOC and that I was very interested in his work with bottlenose dolphin communication. I was especially interested in the implications of his work regarding inter-species communication. The experiences that I had had during the channel islands expeditions, and the reports I had read of the work's of Dr. Herman, Dr. Clark and others convinced me that my own interest in developing a personal and intimate connection with Tomas's "Whale People" was shared, albeit perhaps a bit more scientifically, by a number of very qualified individuals in the field of cetacean studies.

I was delighted then when Dr. Herman invited us to the station in Honolulu to observe a research run with his dolphins. Before the run I met with Dr. Herman and his colleague, Dr. Adam Pack, for an informal discussion while my students visited with his laboratory interns. Realizing that this was a unique opportunity for me, I chose to drop all professional hesitancy and simply expressed my beliefs regarding the sentience and spirit of the cetaceans, and how certain I was that we were only touching the surface of the potential that existed for inter species communication. I was rewarded with not only the attention of these two intelligent and accomplished men, but their whole hearted agreement that indeed, at the very least, these were creatures of great intelligence and that working with them naturally resulted in a high regard and respect for them as sentient creatures far beyond the popular conceptions of both scientific and lay persons. Our discussion became more animated and passionate, and though I suspect that they would hardly remember me now, at that moment, I felt a deep respect and camaraderie with these two men. I gained much from our talk and when we went out to the dolphin tank with my students they, my students, could tell that the energy was very high between us, which in turn, energized them. What I believe had inadvertently happened was that when the session with the dolphins began, a new and powerful energy was radiating from my students and myself. Since we were asked to remain quiet and in a group on the platform above the tank, our energy was concentrated in a ball above the dolphins. One of my students however stood a little bit off from the rest of us and I could tell by his body language that he was literally transfixed. I had been in that state myself so I let him stay where he was! I cautioned my students to do nothing to disturb the research session.

When the session was completed one of the dolphins veered from the other dolphins and swam to a position just in front of the student who was standing alone. The two looked at each other for a long time as though they were passing something between them mentally and then the dolphin raised it's body vertically so it was closer to the student, chattered something to which the student responded with the most delightful and deep laugh. The dolphin leaped, turned and swam back to its mates. When we gathered around Jim he was as if in a daze, all he could do as we all congratulated him, was to smile in a way that touched all of us...it was a truly "beatific" smile. Dr. Herman commented later that it was very rare for one of his dolphins to do that. For the few days left for Jim in Hawaii with us he was a transformed individual, all he could do was smile. For myself, I was convinced that indeed something had transpired between Jim and the dolphin that was outside the parameters of scientifically accepted inter species communication. It was a dramatic way for this voyage to end. I had been on the sea for 6 months, became familiar with the humpback whales, conquered my fear of the sea and was more certain than ever that John Lilly was correct in his assessment of the sophisticated intelligence of "The Cetacean Nation". I was also becoming more comfortable in thinking of these highly sensitive and intelligent creatures as "The Whale People".

For me, and for whale studies in general, the new scientists had opened the door to a more learned and sensitive appreciation of who these Whale People were and what they might represent on our planet. For me also, the journey had really just begun.

Chapter VI
Endings, Beginnings, Endings...

The separation from my third wife during the ATOC expedition resulted in a reconciliation and I returned to our home in the summer of 1999 with the hopes that a new beginning could be had in my so far disappointing marital track record. We traveled together well and thoroughly enjoyed a number of trips together for the next four years. My interest in whales, at least from the research perspective, had waned and I had no real desire to continue with Pelagikos so I resigned my position and became a thorough land lubber again. I assumed once again that any chance of my really becoming involved with "The Whale People" on the level I wanted to was slim at best and my connection to them was over. I was fifty eight years old, lived deep in the mountains of Mendocino County in Northern California and I had had my big experience with the whales, certainly more than most people are afforded. To a certain extent I even felt fulfilled. I was very happy in my teaching position at the college and seemed to be growing more and more into the professorial mold that had begun in 1973 when I was hired as the first Biology/Chemistry instructor at the then brand new Mendocino Community College in Ukiah, California. I now had almost 25 years of classroom experience under my belt and had written a PhD dissertation on higher education and social change entitled "Education as a Ritual Process". I was teaching biology and environmental science, my two favorites, and had abundant opportunity to express, through practice, what I had preached in my dissertation.

In 1999 my wife and I took our second trip to Baja California Sur, this time flying to visit friends who had bought property and built a house in a small fishing village called El Cardonal on The East Cape. I was not particularly enamored with El Cardonal but my wife was taken with it and decided to buy a lot there. The climate was nice, the sea tranquil and clean and there was a small inshore coral reef head that provided good snorkeling and SCUBA diving. Kayaking was also available so I did have things to do. Susan was adamant about returning the next year but this time with a trailer. We found a 1967 Airstream trailer that was in fair shape, and after some work it became real comfortable. We drove it down the Baja peninsula in the year 2000. Susan was also prompted to buy another lot adjacent to the first one so we would have a buffer from further development in the area. I was still not convinced that El Cardonal was such a good place to spend much time, but it was certainly a good investment.

We also made, what was to me, a pilgrimage to Egypt. Ever since I was a child I was fascinated by everything and anything that had to do with Ancient Egypt. I became an immediate "Egypto-phile" and continue to this day unable to put down anything that deals with that incredible and advanced culture. The trip and what I assimilated is a book on it's own, suffice it to say that I have never been more spiritually moved except when I have attended Medicine Ceremonies of The Native People of The Americas. I returned from Egypt with a strongly fortified sense of spirit and purpose, but without direction. Perhaps the fortification was for what would transpire during the next couple of years.

As much as we were trying, our marriage was not working and we separated again, this time for good. In 2002 we called it quits and I once again moved in to a small rental in the area. In the meantime my daughter graduated from Cal State Sacramento and had married, and my son was attending graduate school at San Diego State University after completing a bachelors degree at UC San Diego.

At this point I decided to offer some Eco-Tours to El Cardonal for students and residents of my then hometown of Ukiah, California. I also began contemplating the idea of spending six months in Baja in semi-retirement and six months in Ukiah teaching at the college. If I was very thrifty, I could teach summer school and the fall semester and actually afford to live there for six months. After all I was sixty eight now and could draw Social Security in another year. So...

While conducting the eco-tours in El Cardonal I began to fantasize yet another Jacque Cousteau like adventure. The in shore coral reef head off the point near my (at that time our) property was beginning to show signs of degradation. It is a popular fishing, diving and boating area and had been "over fished" for the past few years due to the increased popularity of the east cape for all kinds of sports and commercial activity. This activity, that is repeated worldwide, has put all the worlds ocean reefs in danger. The Australian Marine Institute had suggested monitoring protocols that could be implemented easily and contribute valuable data to just how serious the situation was becoming. It seemed natural that with the simple protocol and my yearly presence in El Cardonal, I could initiate a study on our reef and contribute to this data collection. I was personally aggravated by the continual shrimp trawling that took place illegally close to the reef and that could not be prevented due to lack of enforcement vessels for the area.

I learned quickly, and to some extent dramatically, how intent were these shrimpers. One day I attempted a Green Peace type action by kayaking in front of a trawler in hopes of disturbing it's illegal activity only to have the shrimper captain increase his speed directly towards me and actually attempt to run me over as I frantically paddled to get out of his way! This experience gave me food for thought indeed, but also impassioned me to actually set up the study.

I had heard there was a young woman, Dawn Pier, in Cabo Pulmo (a major and technically only true hard coral reef in The North Pacific at these latitudes) who was actively engaged in the study and protection of that reef. I arranged to meet her and she inspired me to seriously begin monitoring my reef.

On my first six month stay in El Cardonal in 2004, now divorced and semi-retired from college teaching,

I decided to do a solo kayaking and SCUBA diving tour around the cape region from El Cardonal to Cabo Pulmo then to Cabo San Lucas, Todos Santos, Loreto (a bit farther north) and then home to El Cardonal. I had chosen to go as far as Loreto because I knew there were some great dive spots there and I had heard that there was a Canadian whale researcher that came there every year to study the blue whale population. His specialty was photo-identification and since I had done some of that myself during the Pelagikos years I thought I might meet him while I was there. All I knew was that his name was Richard Sears. Richard and one of the blue whales of that population near Loreto were to become the inspiration for my return to whale work, but this time in a very intimate...and indeed personally spiritual way - just as "The Whale People" had perhaps intended?

My first stop during this odyssey was at Cabo Pulmo, and was also to be my first underwater encounter with a shark. It was a white tip reef shark, so not considered at all dangerous but a real shark nonetheless. All my previous shark experiences had been with leopard sharks while working for the marine lab in Bodega, and they are even less dangerous than the white tip reef shark. It is worth telling the story of this encounter because it too added to my increasingly important connection with the sea and especially my personal comfort while on, in and under it.

I had kayaked out to the second finger of the Cabo Pulmo reef and had spotted a school of barracuda under me so I rolled out of my kayak to get a better look. They were truly beautiful and took little notice of me until suddenly they darted away and I felt the presence of something bigger nearby. I turned to my right and there, not more than 4' away was a 7' shark. The shark was clearly aware of me and as it swam next to me and I got a good look in it's eye I felt such a primal energy in him that I became a bit, but only a bit, frightened. As I was running out of air (I was free diving) I slowly began to rise to the surface, all the while watching him closely because he was watching me closely, and I inadvertently came up right under my kayak and banging my head loudly on the hull. This is when I got my second hit of primal energy as I saw the shark take off like, well, I don't know what to say but like only a shark can. OK, like a flash and he was gone...but where? The visibility in The Sea of Cortez even on the best day is no more than 40-60' and so he was gone and out of my sight - "in a flash"! I don't remember lifting myself into my kayak, I think I must have shot out of the water as if catapulted and found myself sitting on it a bit out of breath and my heart pounding. As I sat in my kayak and chilled out I began to feel ashamed. Here I was, a well read and experienced biologist, who had made a point of telling my students that sharks were not innately aggressive or dangerous yet I had reacted like an unlearned novice. What to do? No one but myself had witnessed this event so I had no one to explain to but...myself. I decided that the only way that I could feel good about the event was if I returned to the water, made myself open to another encounter and wait until whatever fear and trepidation I had, had totally dissipated. I rolled back in, dove down, scanned the area and then returned to the surface. I then turned on my back and lay spread eagled on the surface until the fear was gone. It took a while but eventually I became absolutely relaxed and at ease. I climbed back into my kayak (with a bit more dignity) and returned to shore feeling very empowered. It looked like my trip around the cape had gotten off to a good start!

A few days later I was in Magdalena Bay on the pacific side of the Baja peninsula to get a look at the famous friendly gray whales of that region. I had driven into San Carlos, the main town there, and was gazing out to the bay when a truck pulled up and a young man introduced himself (Gabino) and asked if I was interested in going out to see the whales. He seemed like a nice fellow and we agreed to meet the next morning at 7:00. I told him I could not afford to pay for the panga myself and he said no problem there would be three other people and we would share the cost. The next morning he was there but no other people. He took me to his home and there I visited with his family while he tried to hunt down some other fares. He came back empty handed but he could see my disappointment and generously offered to take me out alone for what I would have paid within a group, generous indeed! His wife had also made us some meat tortillas and it sounded like it would be fun. We found the gray whales near the entrance to the bay and about ten miles out. Just as advertised the whales approached us and I was treated to my first really close encounter with "The Whale people" of Mexico. A panga is a 15-20' outboard motor boat that is used by all fishermen in Mexico. They are very stable, broad beamed and reliable with about two feet of free board so when the whales were close, they were really close. By this time of course I had had a number of whale experiences and I was not as "amazed" as I might have been. It was however a very pleasant day with Gabino - and the tortillas were excellent.

Urmas Kaldveer, PhD

From Magdalena Bay I drove to Loreto and set up camp on Juncalito Beach just south of that charming town. The next day I went in and I asked around to see if anyone knew where "the Canadian whale guy" might be. I was given directions to a house near the Malecon (board walk) but before I got there I saw a group of young people on the street side and thought they might know... my teacher feelers were out and they looked like student interns. I approached the group and asked if they knew Richard Sears. An older (not as old as me however) man among them replied that he was Sears and he invited me to their rented house for lunch. I was immediately taken by this man, and as he told his story, I began to realize that this was one of the current new scientists and totally devoted to his whale research. He was involved primarily in photo-identification of the blue whales around Loreto and had been monitoring that population for nearly twenty years. His enthusiasm, intelligence and respect for the whales touched that unfulfilled spot in me, and my fantasy to see into the "eye of the whale" took fire once more. He was going out on a search the next day but could not take me due to the number of people already going in the panga but said that he had seen a number of blue whales the day before and that I should be on the lookout for them.

I returned to my camp and contemplated this recent event. I had personally met one of the top whale researchers in the world and he was willing to talk about these creatures in a way that hinted to me that he too was aware of the delightful connection that could be made between "The Whale People" and we humans. I owe Richard for inspiring me at that moment to reawaken my dream of working intimately with the great whales. I have been fortunate to have developed a friendship with Richard since then and have told him of the tremendous influence he had on me.

The next morning I awoke to what was to become a beautiful Baja day. The sea was as still as I had ever seen it and there was no wind line visible in the distance. Isla Carmen is just four miles distant from Juncalito Beach and without a second thought I grabbed my gear, some food and launched off the beach in Haldjas, my trusty kayak. I paddled straight for the island and as I got further and further away from the beach and started into the no man's zone (meaning no other humans) the more relaxed and joyful I became. There is something very special about being in a kayak far away from shore and all other human activity, on your own, not responsible for anyone else but yourself....and that responsibility is total. There are no life savers out there, no boats, nada... just you - it is all up to you. As I approached the island I saw some blows ahead of me and they were definitely blue whale blows, narrow, straight and 15-20' high. I counted eight whales and as I got closer (300' or so) they split up into two groups with one going north, the other south.

I figured that it would be best to follow the ones going north so if the wind and/or current did come up later I could kayak with it instead of against it. Following whales in a kayak requires some pretty serious paddling unless they are involved in some social or feeding behavior that keeps them in a concentrated area. The whales I followed seemed to be in feeding mode (inconsistent direction, lots of flukes up and surface lunges) so I was able to get relatively close to them but not right up along side. It was nevertheless a very special, and yet again, new whale experience for me to be within a group of whales in my kayak and for them to be blue whales, the largest creatures our planet has ever known. Being utterly alone there among those whales and in that beautiful place I felt more relaxed, alive and in the moment than I had ever experienced before. I felt their great energy and, for at least a time, I was sharing their, and my, ancient genetic relationship by simply being in each other's presence as joyful spiritual beings. At this point of course I had no scientific evidence of this shared experience but in my heart and my soul I was certain that indeed we were in communion.

I stayed with them for an hour or more and then noticed a wind ripple on the water's surface and since I had a four-mile kayak home ahead of me, I slowly began paddling back to Juncalito beach. I did not want to get caught in the middle of the channel between Isla Carmen and the mainland if the wind or tide became an issue. As I was about half way across I saw to my right (north) that there was a single whale about a mile off and moving at a cruising speed parallel to the mainland and might, just might, be in the right position for a close encounter if I continued in the very direction I was paddling. As the whale continued it's swim our intersect point began to get closer and closer. Each time it dove after the classic three breaths and flukes up I adjusted my speed and direction to maximize the chance for the encounter. By the time I was about one mile from shore I could clearly see the whale as it approached and it was a big blue whale...really big. At a distance that I would estimate as perhaps 100 yards it fluked up and went down and there was no question now that we were right on target for an encounter. By this time I was certain that the whale was keenly aware of my presence and position (their sound and touch sensitivities to the most minor fluctuations in water movement, direction and strength, not to mention temperature salinity etc, are so far beyond our own as to be almost incomprehensible) and seemed to be perfectly in communion with my presence as we drew closer. I paddled a bit further and then all my senses told me I was in the exact right position for the moment it would resurface. And then it rose out of the water like the classic image of a WWII submarine surfacing in an old movie and coming literally right at me from my right and not more than 100' away. I was absolutely and exactly broadside directly in front of him. I realized that if it was not aware of my presence or saw me in any way as a nuisance or threat that I was in trouble. I thought of paddling out of the way, but as with the shark at Cabo Pulmo, I needed to conquer my fear, so I sat and waited. The whale, now appearing enormous, came to within 30' of my kayak, arched, and without causing a ripple, slid gracefully into the sea, passing directly under my kayak at a depth where I could actually make out the markings around it's dorsal fin, resurfaced on the other side about 50' away and continued it's stately journey south. I was awed, I was thankful for the gift and I was filled with a love for this creature that was truly heart expanding. I knew that I had been privileged to be given this moment, and as I kayaked the rest of the way to the beach, the tears began to fall and I realized my time with the whales was definitely not at an end.

I went to see Richard the next day and told him of my experience. He himself had not seen any whales that day and though unhappy that he had missed an opportunity to ID the whales I had come across he was genuinely happy for my experience and could see I was stoked. He then asked me if I might consider ID'ing any blue whales that passed close to El Cardonal the following season (2005). I replied "of course", and without knowing it, this was to be the real beginning of a now eight yearlong journey with "The Whale People". That year, 2004, was also the year that the largest and most intensive whale study ever conducted (SPLASH: Structure of Populations, Levels of Abundance and Status of The Humpbacks of The North Pacific) was to begin in The North Pacific, and Mexico's Dr. Jorge Urban Ramirez was going to be directing the research activities in Mexico's waters. For the next three years, ten nations and over three hundred researchers would be monitoring the humpback whale population of the entire North Pacific, the greatest whale study ever attempted - and photo-identification would be the core of that study. This study was unknown to me at the time, yet in less than a year after my visit with Richard in Loreto, I would be a team member in Dr. Urban's contingent of researchers in The SPLASH project, and my opportunities to see into the "eye of the whale" would literally become a part of my very life.

Chapter VII
SPLASH and The Baja Era

Returning to Ukiah in June of 2004 to teach summer school, I was terribly excited about what the following winter would bring in terms of whale research in Baja. I was closer to realizing my fantasy of seeing into the "Eye of the Whale" than ever before. I knew I needed some funding in order to pay for the rental of a panga/pilot and I thought also that it might be nice to include two volunteer intern positions to aid in my work. I made an offer to the students in all of my summer school and fall semester classes to have two of them join me in Baja in February of 2005 to engage in photo ID work with the blue whales passing through my part of the east cape region. Although I was still interested in doing something on the reef, the whale project took emotional precedence, so all my fund raising activities focused on having the money to rent a panga and the services of a captain or pilot for that purpose. That first venture into fund raising was successful enough to enable me to go out twelve times during the peak of the whale season, mid-February through mid-March. Two young women from my classes at the college volunteered to join me during that period. Lenee Goselin and Kristin Paiva, both from Lakeport, California, became my first two interns in a study that is now in it's eighth year and has born fruit in ways totally unexpected.

During the previous (2004) winter period in El Cardonal, I had asked around in the village about who was a good pangero (operator of a panga) and was told the best man was one Vicente Lucero. I went out with him twice that winter and found him to be a personable, knowledgeable and jovial fellow of about forty five whose family had been fisherman in the area for generations. His English was better than my Spanish so he spoke English and I Spanish, correcting each other when necessary. It did not take long for us to establish a repoire and we soon became friends.

Urmas and Vicente, pilotos, amigos and the dynamic duo of The Sea of Cortez. For the past eight years we have monitored the humpbacks within our area during the breeding and calving season. As of 2012 we have ID'd over 200 individual whales.

When I returned to the east cape in January of 2005 I went directly to Vicente and asked him if he would be my "Capitan" for the duration of the study. He was a bit miffed at first because when he asked what kind of fishing I was looking for I replied, "yo qiuero buscar por

Las Ballenas, no las pescas" (I want to search for whales, not fish). It took him a while to realize I was serious yet it still didn't fit at all into his concept of a gringo's interests in Baja. We made an agreement and I paid him the entire amount of money (supposedly never done in Mexico before the work is done) and he said he would throw in two extra times out as a show of good faith and friendship. To this day, that deal still stands between us, and outside of a time or two when he was "incapacitated" so we couldn't go out; he has been as reliable as the sunrise. I consider him as more than just a friend; he is more like a younger brother. We have had dozens of whale experiences together and both continue to marvel at the beauty and grace of these beings each time we find them.

That first season in 2005 was in many ways exceptional. Lenee and Kristin turned out to be delightful, hardworking and enthusiastic companions. They also brightened Vicente's day with their youth and natural beauty, both being very attractive young ladies. In a short time we became a team and were able to obtain over 40 IDs.

However, we did not see even one blue whale, they were all humpback whales! After the first few times out with no success finding blue whales I decided to ID the humpbacks as well since I knew that someone would be interested in the pictures, and after all they were in my grid in substantial numbers. I felt bad for Richard (Sears) because it was to aid in his research that I had put this project together in the first place. I did not know then that the blue whales tended to pass my area in deeper water about ten to twenty miles out, thereby skirting the grid I had established (in hindsight I find it ludicrous that I missed this because I knew that blue whales were not shallow water whales and therefore stayed further off shore when migrating along the coast). The cost of a panga rental to a great extent goes to fuel so in order to stretch my funds as far as I could during that first season I kept my searches within a grid that allowed maximum coverage for least amount of travel time. This grid consisted of approximately one hundred square miles and took me as far as La Ribera South and Boca del Alamo North and two miles from shore to ten miles from shore. This was an area that could be covered in six hours depending on number of whales and successful photographic encounters.

A "successful" photographic encounter results when one spots a whale, is able to get within 100 or so feet of it and is then able to obtain a picture of the ventral (underside) of the flukes as the whale dives after it's third or fourth blow. This "photo op" moment is usually less than three seconds and is more often only 1 and one half seconds if you want a really world class photo ID.

This is a "world class" ID, full flukes, in focus and with definitive markings. This is "Dottie", adopted by Terry Kerr of Charlotte, NC, a very dear, dear friend.

The photo needs to be in focus and show as much of the underside as possible so that the markings are clearly identifiable by another researcher if "recaptured" (photographed again in another season and/or in another location). A data sheet is also kept for each whale, recording pertinent information. That first season we did not recapture any of the whales we saw during the two months that I was active, but six of my whales were seen by other researchers in different locations in Mexican waters during that time.

Although my deeper quest was to understand these creatures in a more intimate and spiritual context, I found that my scientific training and my personal feelings of obligatory professionalism in the field dominated my activities and I therefore pursued the work with scientific focus. Nevertheless, being in such close proximity to whales, and for an extended time as I was, gave me ample opportunities to experience the eye of the whale. As I mentioned before, we did obtain over thirty good, and perhaps twelve world class IDs that first season and obtained valuable data for my area. When the season was over I had to tell Richard of my failure to obtain any blue whale IDs for him and asked him who might be interested locally in receiving my humpback photo IDs. He suggested I contact Dr. Jorge Urban Ramirez at The University in La Paz (Universidad Autonoma Baja California Sur, UABCS) and that is what I did. Here is my first correspondence with Dr. Urban. I include it in this text because it represents a commitment that overnight, placed me in a position to seriously contribute to whale research.

Dr. Jorge Urban
Dept. of Marine Sciences
University of LaPaz
LaPaz, BCS, Mexico   June 27, 2005

Dear Dr. Urban,

As a semi-retired college professor from California, and having worked with both Blue Whale and Humpback photo identification in the past, I was intrigued by the number of Humpbacks I saw this year near my home in El Cardonal on the East Cape.

I could not resist the temptation to attempt some photo ID's. After shooting 350 photos I have between 20-30 that I feel are verifiable. I also kept records of all details of the encounters (GPS, behavior, calves etc).

My friend Richard Sears suggested I see if you would like to have copies of these photos and data. I would be delighted to share them with you and to continue the work next year with your approval.

From what I saw this year, I am personally convinced that this is an important calving and breeding area. The area is also becoming increasingly popular for sports fisherman and recreationists. When I was tracking Humpbacks in Hawaii in 1998-9 we found that small engine noise was the most detrimental to calving and breeding. To our surprise jet skis were the most offensive to the whales. Therefore it is very possible that the area I speak of will become more and more difficult for the whales to use.

If my work can be used by you to help people understand the life and needs of these great creatures I am at your service.

Urmas Kaldveer, Ph.D.

Dr. Urban replied immediately and encouraged me to keep up the work the following (2006) season. He also asked if I would like to join his team.

What I did not know was that being included on Jorge's team meant that I had become a member of Mexico's contingent of contributors in the largest whale study ever attempted...SPLASH. It would still be necessary for me to raise all my own funds but I was now involved in a project that could significantly alter our way of thinking regarding the humpback whales of the entire North Pacific Ocean. For me that implied that there was therefore a tangible scientific goal and value to my work.

Dr. Jorge Urban Ramirez: Professor and Coordinator for marine mammal research at The Universidad Autonoma de Baja California Sur in La Paz, Mexico and more importantly to me, a friend.

Regardless of my immersion into the science of whale lore that first season of 2005, it also provided ample opportunities for me to become more intimate with these creatures and to continue to inspire me towards my spiritual quest. Two events that first season stand out that I would like to share.

The first took place early in that season and involved the first mother (I can not abide using the term cow for a mother whale) and calf pair that I had ever encountered. We were about one mile off shore from Punta Pescadero in the southern half of my grid and "the girls" (they accepted this term as one of endearment) spotted a large humpback moving slowly north. I had been training Vicente in how to make a non-intrusive approach on the whales and since this was our first mother/calf encounter I wanted to be especially sensitive to the mother's energy (in Hawaiian Waters, where there is thought to be the largest breeding/calving population of humpbacks in the north pacific, there was already a strictly enforced approach law of no closer than 300'). To our surprise, and delight, it wasn't we that were looking for an encounter but the whales. The calf was quite young, most likely a yearling born in Mexican waters the previous season, and the mother very large probably indicating it was not her first calf. She circled the panga (24') with her calf a number of times (I asked Vicente to cut the engine) and then allowed the calf to approach us more closely. The calf not only came close but actually brushed against the side of the panga and blew a fine mist of seawater and "bad breath" on us in the boat (this is often the case when observing gray whales on the Pacific side but very rare with humpbacks and almost unknown with blue whales). I was able to look directly into it's eye from only three feet away as it passed and was struck by the level of awareness, curiosity and intelligence that was manifested in that glance.

This would be the first time I was to have this particular experience so close. Not only was it literally looking into the eye of the whale, it was also a taste of what I was to experience many times again as a part of my personal quest for communion. Indeed, since 2005 I have had numerous eyeball to eyeball encounters with humpbacks from the panga, from my kayak and often under the surface with them as we swam together. More than half of those times it has been with a mother and calf pair seeking encounters with me out of what I can only describe as curiosity and friendship - more on that later.

The other story I wish to relate about that first exceptional year occurred the last day we went out for the season. Lenee had become increasingly adamant about wanting to swim near a whale. As she was my responsibility I had to tell her no, even though in hindsight it would have been perfectly safe. What made things different that day was that after tracking a humpback for a while and getting a good photo ID the whale swam under the panga and began to sing. This was a new experience for me and it was thrilling to not only hear that ancient and intelligent voice but the sound vibrated the bottom of the panga so our entire bodies experienced the song. Humpback whale song is far more complex and intricate than most people are aware. I will present some recent research in the next chapter that I believe will be surprising to most everyone. This was all too much for Lenee, she threw off her shirt (we always had suits under our outer gear just in case) and shorts and without asking simply dove over the side and threaded water, diving deeper every so often to hear the song better. This was too much for Kristin, so she did the same. It was a delight to see these two young students who had been so scientifically involved with the project totally letting go and entering into "whale space".

The whale moved on, the girls climbed out and I realized that they had taught me something about following your heart. I on the other hand I was sixty four years old, certainly too old to start swimming with whales, no?

Returning once again to Ukiah in June after that first season, I immediately started raising funds for the 2006 season. I had come to recognize that it was very comfortable to divide my year in half, teaching part time at the college and spending my winters in Baja tracking the humpbacks. All my whale experiences thus far had strongly reinforced my personal opinion, shared by many, that these creatures were far beyond the level of intelligence and sensitivity we gave them credit for. By this time I had shared my pictures with Jorge and had been invited to be on his team but I knew I needed a better camera. I was able to raise enough funds to purchase a fine 35mm digital Minolta/Konica Maxim 5 with a 280mm telephoto lens that I am still using today.

2006 was another good year but there were fewer whales so I only obtained about a dozen good IDs and only one or two of those were world class. The lower numbers were disappointing but did not reflect anything serious in regard to the well being of the humpbacks in my area.

During the fall semester of 2006 one of my students had brought in a video made by a friend of hers who turned out to be one of the drummers for *"The Grateful Dead"*. It was called *"Ocean Spirit"* and it recorded a voyage taken by him and four of his friends into the north pacific specifically to swim with as many ocean critters as they could find. Watching these men swim with whales, sharks, rays, sea lions etc. I was inspired! Humans, free diving with all manner of critters, fearless, full of respectful loving energy and I was hooked. How can you look into the eye of the whale if you don't enter it's world? I vowed to myself that the next season, 2007, I would swim with the whales.

2007 was another very good season with lots of humpbacks, and much to my, and Richards delight, I was also able to ID twelve blue whales along with my humpbacks.

This is one of the first blue whales that I photographed that year. We were able to stay nearby it for three blow series and were treated to what I consider to be one of the most beautiful sights in nature. The majestic grace and power of this "critter" literally took my breath away.

On March 17th of 2007 while tracking a humpback on a most beautiful day I decided to make my first whale dive. At first Vicente was very skeptical about my doing this because he was convinced that there were man eating sharks everywhere in The Sea of Cortez and I was foolish to voluntarily enter the water and tempt them. My previous dive in Cabo Pulmo in 2004 with the white finned reef shark and my successful desensitizing of the fear that went along with that experience assured me that all would be well. I put on my gear and sat on the gunwale waiting to bail out as soon as Vicente could maneuver the boat into a position in which I could get in the water in front of the whale and wait for it to arrive. Vicente skillfully put me in position, I bailed out and was able to get within about 60' of it before it "fluked up" and dove too deep and fast for me to follow. On board was my colleague and good friend from the college, Susan Janssen, and she took pictures of this first dive.

Shortly after this first dive with a whale, Vicente and I came across a mother and calf pair that was cruising south near Punta Pescadero. The mother would not fluke up due to the youth of the calf and not wanting to leave it near the surface without supervision. I took flank shots of the pair and noticed that the calf did not look at all healthy. I wanted a closer look but did not want the boat to frighten the calf or anger the mother. I asked Vicente to position me well in front of them and then I dove in to wait. The timing and position were perfect with both whales diving just before arriving at where I waited in the water. I then dove hoping to see the whales as they passed nearby and get a close-up look of the calf's body. I didn't see the mother but I got a good look at the calf. I was both thrilled (this was my first underwater encounter) and dismayed. The calf was only a few feet away and was clearly curious as to what I was. Its eye was on me and it turned its body slightly to get a better look at me and I felt a strong connection with this intelligent little critter. The calf however was in very poor health; the backbone was pronounced and the skin looked pasty and gray. There seemed to be a number of open sores and a fair number of long lacerations across its body. I feared that it had been badly entangled in a drift net or gill net. The marks did not look like Orca rakes. I have had a number of underwater encounters since then and it is often the case that a calf will be the initiator of closer contact.

It was exhilarating and very empowering. I now understood the joy of diving with the critters that was so well documented in *"Ocean Spirit"*. Though there was no interest on the part of that whale in making close contact I was certain it was fully aware of my presence and I felt honored to be in it's space. I made eight more whale dives that season and each one, just like the ones I continue to make now, is a special treat in my life. I have yet to have an adult whale stop and visit but I have swum alongside many both above and below the surface and have had the truly wonderful experience of having five whales make a point of swimming to me and letting me look into their eyes from a distance of less than 20 feet. In April of 2010 I had the very, very special experience of diving with a blue whale close enough so that I was able to look into it's eye and believe me when I say, that was a treat. My Native American friends in Northern California refer to the blue whales as the ocean's "Wisdom Keepers". Once again my dear friend Susan Janssen was on board, how sweet!

I had also begun kayaking out to what I referred to as "The Whale Zone" (WZ) every few days to see if I might have an encounter. The WZ is two miles out from shore directly in front of my village (I also refer to this area from one mile to five miles out as the humpback highway). I go out early in the morning and usually stay out for about two hours. I have had many wonderful experiences out there. I have swum with five different species of whale (including a pod of twelve Orcas), four different species of dolphin, three species of rays and a number of sea lions and turtles. Never once have I felt anything but acceptance, interest and an awe at my good fortune to be with these critters in their world.

As a way of raising funds for each season in 2008 I started an "adopt a whale" program. 2008 was special in that regard in that my son Zack adopted a whale that year which he named "Odin" (Zack is very proud of his Viking heritage, as am I). I recaptured Odin six times during the 2009 season and once again during the 2011 and the 2012 seasons. Odin has returned to my grid three times after being here the first time and he is the only recapture I have experienced to this date. A recapture is when an individual whale is photographed during two separate seasons (do not to need to be consecutive seasons) or photographed in both its southern breeding/calving area as well as its northern feeding area.

2008 and 2009 were also somewhat difficult years. In 2008 my doctor informed me that my PSA tests were showing an increasing predilection for prostate cancer and that I should have a biopsy taken. From the stress of that possibility I developed a case of Shingles and suddenly found myself seriously concerned for my health. This was a first for me because all my life I had been blessed with good health. I made it through the Shingles episode with far less discomfort than most other victims but it was a bad experience nevertheless. I continued to fret about the possibility of cancer and read a great deal of literature about PSA tests, biopsies etc. I decided that I would not have a biopsy and would instead use all my personal spiritual powers to heal myself. I began to eat better, meditate every day and in general put my life into better order. I also decided to end my thirty two years of teaching and move to El Cardinal, Mexico fulltime. I calculated my financial resources and determined that it could be done as long as I lived very frugally and Social Security remained viable. In January of 2009 I packed up all my worldly goods (they all fit in my Subaru Impreza) and did my last drive down "The Baja" to my new home in Mexico.

I was sixty eight years old and ready to confront my mortality in the village of El Cardinal where I had become to feel at home in a way that I hadn't felt in fifteen years. By this time I had also completed four years of collaboration with Dr. Urban and had begun to feel that The Sea of Cortez and its critters were part of my family. That first fulltime year (2009) here in Mexico had two more unwelcome surprises; I passed a kidney stone (again fortunately without the terrible pain I have heard about, but enough to bring me to my knees nevertheless) and a CAT Scan showed a possibility of a tumor in my other kidney. My skin also showed signs of a possible melanoma cancer (I spend a great deal of time in the sun). All this successfully put me in a morose state bordering on deep depression and at times even suicidal. The kidney tumor turned out to be a cyst, I healed my skin with Burt's Bees rescue remedy, I began drinking Creosote Bush Tea (suggested by a local Curandero and collected locally in the desert) for both future kidney problems and possible prostate cancer and began walking "The Medicine Wheel" as taught me by my friend "Tomas" from Killarney. I am certain that these efforts on my part to a great extent healed me and turned my depression into an avenue for the rejuvenation of my spirit...and thereby my health. I was also helped immensely by Dr. Ron Scolastico, a Psychic who introduced me to concepts that brought yet another method of healing into my repertoire and Louise Hay, whose words literally sprung me into positive self healing action.

A great addition to my healing was an exceptionally good year of whale activity in my grid during the 2010 season with many encounters and memorable whale dives. Vicente and I ID'd eighty-three individual humpbacks with a good thirty being of world class quality. I was beginning to feel as though the whales had become family and that my presence in the water with them was a form of communion that we both enjoyed and appreciated. It was also during that season that I had a very special experience while kayaking in The WZ.

I was out at my usual two-mile mark, and though my spirits had risen, I had once again developed a health problem and found myself ready to give up. It was a flat and beautiful day, warm and delicious in every way...except for my feelings of defeat. About 30' away I spotted a sea turtle (a Green Sea Turtle I believe) on the surface and decided to see if I could approach it before it dove away. All of my turtle experiences to that date resulted in the turtle diving when I was generally within 10' or so. I kayaked very slowly toward it and was delighted to see that it wasn't diving. The closer I paddled the more excited I became. The turtle looked toward me, snorted a few times but let me paddle right up next to it. It was about two to three feet in length and allowed me to get right alongside it.

Urmas Kaldveer, PhD

As it swam around my kayak it seemed to be interested in it and kept bumping it with its head and carapace. At one point I reached over and stroked its carapace fully expecting it to dive...it didn't, it just kept looking at me over its shoulder. I then decided to stroke its neck, which turned out to be smooth as velvet and very soft. Still it just continued looking at me. By this time I knew this was a very special moment and I thanked The Great Mystery (the native Huichol designation for an over soul) for this blessing. After a bit the turtle began to swim away but continued looking over it's shoulder as if to see if I would follow. I rolled out of my kayak thinking that perhaps it would let me close and I could swim with it for a while. When I got within about 3' the turtle dove and I dove with it. As the turtle descended into the darker water it still kept looking back at me and I could not help but think it was beckoning. Then in a flash of awareness I realized that perhaps my time had come. If I were to continue following it (the turtle), maybe it would take me away from all my earthly problems. In other words I thought perhaps this would be a good way to end things. The year before I had cleared with my son and daughter that if I ever became physically incapacitated to the extent I could no longer engage with the sea and the whales that I would chose to terminate myself. I was terribly proud and gratified when they both said they understood and were behind me 100%. I cannot begin to articulate the sense of freedom they gave me and that I cherish to this day.

By this time the air in my lungs was just about finished and I thought, "what a way to go"! Suddenly out of nowhere a bottlenose dolphin jets from my right side, crosses between myself and the turtle, looks me sternly in the eye, and I hear a voice in my head that say's, "no, not yet, you have things to do", and then it was gone. I kicked hard and returned to the surface gasping for breath just like in the movies. Climbing back into my kayak, somewhat chagrinned, I contemplated the event and indeed interpreted it to mean just what I heard, "no, not yet, you have things to do". But what was that?.

Well, I gave a power point lecture in San Jose del Cabo that year where I proposed an idea. **One Planet: Two Worlds**. I will explain this idea fully in the next chapter. What did answer to some extent the question as to what I was yet to accomplish, was provided by Jorge when he told me that it was very possible that his work (and to a small extent mine) had convinced The Mexican Authorities to consider designating the cape region of Baja Sur as a marine mammal protected area (MMPA).

Another event occurred during the 2011 season when I began to wonder if I hadn't become a bit too self aggrandizing in my whale stories, particularly regarding diving with them. I had already been extremely fortunate in the intimate experiences I had had thus far. I thought that perhaps I should refocus on the scientific side of my work here in The Sea of Cortez and be happy with what I had already accomplished in my personal/spiritual quest. One morning I awoke to a deep feeling of need to address this question (which had become a moral issue for me) about my swimming with whales...was I approaching these wonderful creatures with my original intent (seeing into "The Eye of The Whale", remember the metaphor) or was I turning it into a stage for my own glorification? I walked The Medicine Wheel I had built in my garden and sat for a long time contemplating the issue. Suddenly I felt an over-whelming need to go out to The WZ and confront the spirits there for advice or….a sign. I looked to the sea and there was no wind so I grabbed my gear and within minutes was paddling out to my usual station.

When I arrived there (2 miles out) I faced each of the Cardinal directions and asked The Niericas (guardians of the doorways to the information held within the four Sacred Directions) to help me access the information and/or understanding I needed. Then I simply sat in my kayak and waited. After about an hour I thought I might start back in because the wind was starting to come up and I still had two miles to kayak home.

On days when I go out to The WZ and there is no action I like to swim laps pulling my kayak behind me (it is always tethered to my ankle) as I get within a half-mile or so of the beach. I rolled out, put on my mask/snorkel but no fins, and began to swim. I know that after a hundred strokes I tend to veer to the right so at each hundredth stroke interval I glance up and correct my vector. I did that at the first hundred and the second hundred but at fifty strokes of my third hundred I decided to surface ahead of time. I had never done this before. When I glanced up, and much to my surprise and delight, there were two humpbacks bearing down on me directly to my right - it was a clear collision course. I knew they were aware of me in every detail right down to my disposition (to be explained in the next chapter) and would not collide with me. When they were within about 30' from me the larger one dove and I dove with it in an attempt to see it under the surface. Again visibility prevented me from doing so but as I resurfaced I saw the other one dive even closer to me and once again I went down. I looked all around and suddenly there it was, not 10' away from me, looking intently at me and we swam together for a few beautiful moments before my air gave out and I surfaced once again. They both rose again to blow on the other side of my kayak and I watched them continue their journey.

Thinking about this on the rest of the way in I wondered what were the mathematical chances that I would choose to go out to The WZ just then, to return when I did, decide to swim in, come up at fifty strokes not one hundred, and have two humpback whales on a direct collision course 30' away from me at just that moment! I am well aware of synchronicity, and though this may have been simply synchronicity, that too is a function of the "oneness" of The Universe. Two other kayakers (my good friends Ann Marie and Dayton) had seen the whole event and said they thought I had seen the whales from afar (they had) and that I had chosen to go into the water to initiate an intercept. They thought it was uncanny how perfect my calculations had been. No such thing, I had no idea there were whales in the area! For me this was sign enough, it was OK to tell the stories, it was OK to attempt communion, if I didn't they would. Was this a continuation of "The Whale People's" Agenda? And what was to be my part in that agenda?

Chapter VIII
Clarity and purpose

By the end of the 2011 season I had had three additional close encounters (within 10'-15') in my kayak with humpback whales, and had seen my old friend "Odin" for a third time and swam with him for a second time. That too was special because once again it was my son Zack's whale that returned here to my grid. I had also been informed by Dr. Urban that his twenty years of devoted work (along with my seven year contribution) had convinced The Mexican Government to propose that most all of the East Cape Region qualified to become a Marine Mammal Protected Area (MMPA). This would include a good third of my grid, and it was that part of my grid, where I had ID'd a majority of my whales. The scientific aspect of my work with the whales therefore had been rewarded with the satisfaction that I had helped protect the humpback whales of the east pacific. In a way it was a personal thank you from me to that very first humpback I came across in my life, and who had for all intents and purposes, saved it forty years before at Bodega Bay in northern California.

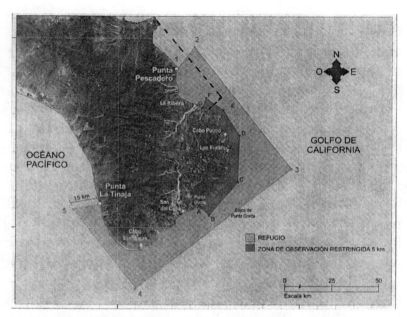

The shaded zone indicates the area of the east cape of Baja California Sur that will be the marine mammal protected area. The stippled zone is the part included in my grid.

2011 also clarified my personal non-scientific quest to see into the eye of the whale. It was as if one day I understood the message I was getting from the whales all these years and now I knew how I was going to deal with the information/insights that they had given me. The first thing to do, when the revelation hit, was to incorporate the insights into my fund raising lectures. This I did by focusing on the core of my revelation, **"yes, we are one planet and all things are connected, but we are two distinct worlds"**, each with their own very special and very, very different physical parameters within which The Universal Life Force has manifested. Both carbon based of course, both questing for the energy necessary to propagate themselves, both following the planetary laws of natural selection...and, in both having followed the dictates of these planetary parameters, have evolved higher consciousness in those animal forms that were fortunate enough to develop the anatomical/physiological characteristics allowing them to manifest the benefits of such development. In the biological sciences we call that parallel evolution (to distinct species separated genetically by long periods of time, having a common ancestor, in this case early terrestrial mammals, yet developing a similar if not identical trait, in this case a complex brain). In the terrestrial world, it has been we humans that have benefited most from the quest for consciousness. Within the marine world it is the cetaceans and particularly the great whales that have benefited most. The conclusion I came to, was that we continue to relate to the great whales as truly beautiful and intelligent creatures but without also recognizing that they are the **OTHER** truly sentient species on this planet. They, just like ourselves, have evolved to develop this intelligence to the higher levels of consciousness we call cognizance, that is, awareness of self - the classic "I think, therefore I am". With that common evolutionary thread that results in the "mammalian brain" (as described by Carl Sagan in his book, *"The Garden's of Eden"*), love, hope, anger and all the other emotions that we consider as being only within the human behavioral realm, are also shared with the cetaceans. We are dead wrong in not recognizing the full implications of this commonality.

The great whales, having evolved in a world so alien to ours, and with such an earlier, longer journey to consciousness, have by the nature of their physical environment (dense, cold, dark, fluid), adapted to a different set of guiding principles as to how this higher intelligence would manifest. My scientific discipline prevents me from using "new age" rhetoric regarding the differences in how each evolving higher intelligence on this planet has chosen to experience their existence, but suffice it to say that all of my thirty years of contact with "The Whale People" has convinced me that their intelligence is as advanced as our own, it is simply manifested differently. People ask, "well, why haven't they done something with this intelligence? Why are there no underwater cities, no art, no...war?" We might then reflect on how we here on the terrestrial world have manifested our advanced intelligence! The important thing is to not apply the same limited boundaries for the expression of intelligence upon the marine world but to see that they are simply different and without applicable, moral judgments to be made as to which **WORLD** is the most advanced. And what pray tell does that mean - advanced?

Before delving into that esoteric and thorny question let me provide the scientific basis for the beliefs that I have described above (the one planet two worlds concept with the cetaceans representing the **OTHER** advanced intelligence on the planet).

Of paramount importance here are a number of basic things to understand: one, that indeed, taken as separate geographical/geophysical realms of our planet earth (yet connected intimately in regard to the cycles that generate and maintain the life force = biosphere), the **TWO**

WORLDS are astonishingly different in regard to the physical demands made on an intelligent creature and the limits to the expression (not inherent capabilities) of that intelligence imposed by those different demands. It might be worthwhile at this point to remember that the mammals that became the cetaceans (whales, dolphins and porpoises), returned to the marine world after having already evolved successfully as a terrestrial animal. Not dominant like the dinosaurs, as stated in an earlier chapter (The Dinosaur Era), but successful nonetheless. In a very real sense a return that required a genetic/behavioral relearning of how to survive in an aquatic environment. That alone was a change demanding dramatic anatomical/physiological evolution but without losing the highly successful evolutionary characteristics already developed by their mammalian ancestors. Included in this would be keen eye sight, the ability to breath air and extract oxygen from it by means of internal lungs and to have internal insemination with a concomitant internal gestation period that maximized the success of a live birth. Not to mention what neurophysiologists refer to as "The Mammalian Brain" (that part that is responsible for the higher thought processes), that began it's evolution towards cognitive capabilities within these clever and increasingly successful land animals, and would one day find it's current broadest expression among terrestrial animals in <u>Homo</u> <u>sapiens</u>. All this gave those mammals that returned to the oceans a very different set of capabilities than existed in any other creatures of the seas.

At the same time it required a genetic reworking of the land mammal characteristics that were no longer beneficial to the prime directive - eat, reproduce and survive. Intelligence however, had already made its mark on land as an extremely valuable characteristic among mammals particularly and was not about to be lost in this transition back to a water environment. At the time of this transition (fifty-sixty million years ago) the terrestrial journey towards higher intelligence, though begun, had another forty five million years to go before finding expression among our earliest primate ancestors, the great apes, and our first distant cousins the Australopithecines (the predecessors of Homo erectus) who began their journey on land a mere five million years ago. It is suggested that the greater the environmental demand, the faster the need for evolutionary change. The change demanded of land mammals returning to the seas could not have been more challenging. Did that intense challenge accelerate the development of cognitive thinking in the cetaceans long, long before it became a factor in our own cognitive journey? Certainly those toothed whales (now represented by a number of species including the sperm whale and the Orca) whose fossils are evidence of a very early and successful adaptation to their environment and are now considered by many as the most intelligent of the present whale species were well on their way to intelligence. By thirty million years ago the baleen whales (whales that graze on large concentrations of plankton and smaller fishes and are toothless) had evolved and added a new dimension to the oceanic survival behavior of the great whales.

What is of critical importance in understanding why many researchers think of the great whales as having this highly developed intelligence is that in order to manifest higher intelligence, we are quite certain necessitates the presence of a specific neuronal cell type called a spindle cell.

Thus far only four animal species on the planet have been shown to have these cells within the gray matter of their brain tissue; whales, humans, great apes and elephants, in order of highest numbers of spindle cells in the brain. This is a cell type that has been shown to provide the rapid and versatile physio-chemical characteristics necessary for cognitive thought processing in the brain. Therefore, barring the innumerable scientific and anecdotal observations of higher intelligence so often witnessed in the field, there is clear neurological evidence to indicate a high order of advanced brain activity among the cetaceans. One of the basic laws of natural selection is that if a genetically determined characteristic does not have survival value, or is not used, it will atrophy and become vestigial or lost entirely. This is especially the case with brain tissue due to it's ultimate responsibility of keeping the animal aware of it's surroundings thereby giving the animal the capacity to evaluate a situation and determine the best order of action. There is little room for genetically faulty or unnecessary tissue in the brain. This is also especially the case due to the enormous amount of blood flow and energy needed to support an advanced brain. Therefore, it is clear that the cetacean brain, which has these all-important cells in great abundance (even more than a human being), most probably is using them for higher cognitive functions. A large number of behavioral characteristics are available without these cells of course, but in order to manifest higher intelligence spindle cells are thought to be required. Before I leave the evidence for sophisticated cognitive processing in the cetacean brain I would like to add that the human brain and the whale brain, though the whales is larger, are almost identical in shape, internal anatomy and surface features (particularly the all important convolutions on the surface).

In order for the potential of a cognitively capable brain to manifest higher intellectual behavior the brain needs to be stimulated by interaction (communion) with other members of it's species whether that is in the strict sense of sharing survival value information or simple social interaction such as love and play. It has been shown conclusively that these are necessary stimuli for the potentializing of neural circuits within the brain. Communication then, or communion, seems to be a very important part of the development of higher brain functions. Almost everyone in today's world is familiar with humpback whale song (the intricacies of which will be described in a moment) but are unaware of the vast repertoire of sounds that are made by the cetaceans constantly in communicating the everyday interactions that they experience.

For a great long time these sounds were thought of as being somewhat limited and represented a comparatively meager set of signals for the basic needs of social interaction. The work of Dr. Rebecca Dunlop of The University of Queensland has shown however that we have been hasty in our evaluations regarding the limits of sound production by the cetaceans. Part of the problem was that we were measuring sound receiving and emitting frequencies only within the human range of 20hz to 20khz. Dr. Dunlop found that cetaceans were communicating with each other at frequencies ranging from 10hz to 110khz, an obviously far greater range than humans. As she herself said, "I've found they have a massive repertoire. I think their communication system is a lot more complicated than we gave them credit for." It appears from her research then that the cetaceans certainly have the capability in terms of sound making capacity to have, and most likely do, a vast communications ability which they are exercising daily in their world. It is critical once again to not limit our thinking to human experience and material expression in a terrestrial world for describing what the cetaceans have accomplished in the evolution and use of their evolved intelligence. What is undeniable however is that communication, in the form of language, is the greatest perpetrator of higher intelligence and learning in all animals and the cetaceans are utilizing this sound making capacity to communicate on a level far more sophisticated than we thought possible for them. And as Chris Clark has said, "A whale's consciousness and sense of self is based on sound, not sight". The key words here are consciousness and sense of self. With their capacity for sophisticated communication, a medium (the ocean waters) where the sounds can be projected for many, many miles (at times thousands of miles) and the advantage of possessing a brain that has the all important, and rare spindle neuron, it becomes abundantly clear that these creatures are not only intelligent, but possess a higher intelligence that we have not even begun to appreciate. They are the most advanced intelligence in the marine world on this planet and their lives are filled with those very same attributes that we consider so special about ourselves...they know themselves and they feel the agonies and ecstasies of being truly sentient.

Before leaving sound I would like to mention just a little about humpback song. I have been very fortunate in having heard a singer twice during the past eight years of research here in the waters of The Sea of Cortez. As I have mentioned, I monitor the migration of humpback whales into and out of my grid here in the east cape. This affords me hours and hours of close and sometimes intimate contact with them, yet in all that time I have been gifted with whale song only twice. If I had a hydrophone set up, that would of course not be the case, but barring that, I feel privileged to have been in the presence of and able to hear a humpback singer. The first time was as described in an earlier chapter when talking about my two interns during the 2005 season. The second time was the 2010 season and as before it vibrated the hull of the panga and could also be heard at the surface. Understanding to some degree, the significance of these songs has made me feel especially blessed to have heard them even once. They are indeed "other worldly", yet initiate very familiar feelings within myself. As Roger Payne described them in 1971 as "exuberant, uninterrupted rivers of sound with long repeated themes, each song lasting up to 30 minutes and sung by an entire group of male humpbacks at once". The songs would be varied slightly between each breeding season, with a few new phrases added on and a few others dropped. Today, humpback whale song is considered to be one of the most complex, non-human, acoustic displays in the animal kingdom.

Urmas Kaldveer, PhD

The following is an article that describes what is considered our present knowledge regarding humpback whale song. It can be found on the **"Whale Trust"** website and their contact information is:

Contact Whale Trust
PO Box 243, Makawao, HI 96768, USAtel 808.572.5700 fax 808.572.5701

Male humpback whales emit a complex series of loud sounds over and over known as song. Singing is usually heard during the breeding season, but also occurs during migration and in late summer and fall feeding areas. Although likely heard by sailors for centuries, the first recordings of humpback songs were made via U.S. Navy ships in the late 1950s off Hawaii and Bermuda. Scientists first recognized these sounds as coming from humpback whales in the 1960s, and the first technical description of humpback whale song was published in Science in 1971 by Dr. Roger Payne and Scott McVay.

Since then, the general structure of the song as well as the basic characteristics of singers have been described by scientists. These characteristics, combined with limited observations of singing whales have led to several ideas as to the function or role of the humpback whale song on the breeding grounds. Our research tests these ideas through intensive observation and measurement of the behavior of singers, and their interactions with other whales.

Structure of the Song
From: Payne, R.S. and S. McVay. 1971. Songs of humpback whales. Science 173:587-597.

The song has a predictable structure with a series of sounds (units), repeated over time in patterns (phrases), with each phrase repeated several times to comprise a "theme". A typical song is then made up of 5-7 themes that are usually repeated in a sequential order. A song typically lasts 8-15 minutes (although it may range from 5-30 minutes), and then is repeated over and over in a song session that may last several hours. The sounds that comprise a humpback song are varied and can range from high-pitched squeaks to lower frequency roars and ratchets.

Humpback Whale Song is Constantly Changing

A striking feature of the song is that it gradually changes or evolves over time. Each year, different sounds and arrangements of sounds form to create new phrases or themes. These changes are slowly incorporated into the song, while some older patterns are lost completely. The song apparently changes as it is being sung. The change in the song display seems to occur in a collective or common way throughout the population. Usually after a period of several years, the song is virtually unrecognizable from the original version. In some cases, however, the song has completely changed in just two years!

Despite the constantly changing nature of the song, all singers in a population sing essentially the same version at any one time. In fact, all the singers in the North

Pacific (that is, whales in Japan, Hawaii, Mexico and the Philippines) separated by thousands of kilometers sing essentially the same version of a song at any one time. The songs of humpbacks may be similar across entire ocean basins like the North Pacific Ocean, yet different in separate oceans. Humpback whales in the South Pacific, for instance, have a different song from humpback whales in the North Pacific Ocean. The explanation for the collective change of the song, especially over such vast distances, is currently unknown.

Characteristics of Singers

Humpback whales sing primarily in winter breeding grounds, although some singing begins in the late summer on the feeding grounds. Singing is also regularly heard during the migration from summer to winter grounds. The evidence to date indicates that only male humpback whales sing. Photographs of the genital region and DNA sampling taken from skin samples have determined the gender of singers. Singers are usually, but not always, alone. At times, male singers may also accompany a female with a calf. Often the singer is relatively stationary, staying in one geographical location for extended periods of time. During these stationary periods, singing humpback whales often hang in an upside down, head-down, tail-up position, approximately 15-30 meters below the surface. The singer will maintain this position until it surfaces to breathe and then immediately resumes this position after diving. At other times, however, singers will steadily travel while singing and may move tens, if not hundreds of kilometers during a song session.

Interactions with Other Whales

Singers will sing until one of two things happen:

(1) they are joined by another lone adult male called a joiner;
(2) the singer stops singing without any close approach by another male and then rushes to approach or join a passing group of whales, often with a potentially breeding female included within the group.

When a humpback whale singer is approached and joined by another lone male, the interactions are usually relatively short ranging from a single pass to rolling, tail lobs, or breaches by one or both animals. Often the pair splits again after a few minutes. One or the other may start singing again shortly after the interaction.

Why Do Humpback Whales Sing?

Why whales sing is an intriguing and complex question. There are several ideas or hypotheses about the social function of song. The dominant hypotheses have viewed the song as a sexual display; with much emphasis on the notion that the song is primarily a display to attract females and that the song changes as a result of female choice. Often attached to this is the idea that the song also serves notice that the singer is prepared to fight. However, our research strongly suggests that the songs function to facilitate social interactions between adult males, with no evidence of a female response to date. Other suggestions on the function(s) of humpback whale song include:

1) a physiological inducement to synchronize estrus in females; 2) or as an orientational display that acts as a migratory beacon, a male spacing mechanism, and/or a form of sonar to locate females. Conservatively, the song is a communication from male humpbacks during the breeding season. It almost certainly provides the location of the singer, and by association the entire herd, and signals that reproductive activity is underway. The song also likely broadcasts information about the individual singer, but what information is communicated and who the recipient is remains unknown.

Portions of the above have been excerpted from "The Song" by Jim Darling in Encyclopedia of Marine Mammals (edited by W.F. Perrin, B. Würsig, and H.G.M. Thewissen, 2002)

Humpback whale song, like the other higher intelligence behaviors of the great whales and cetaceans in general once again suggests a creature that is highly evolved and could very well be considered a second intelligence on our planet. Isn't it interesting to contemplate a future alien discovery of our planet, but instead of finding the terrestrial intelligence worth study, they would choose the cetaceans?

Beyond the fascinating implications of their sound making that we have recorded, there is also a vast record of surface behavior that is worth noting. At present there are at least sixty-four separate surface behaviors that have been identified within the humpback species and shared by most great whales. All of these behaviors have been documented over long periods of time and have been shown to be, not just another means of communication, but are often indicators of curiosity, and perhaps most telling of higher intelligence...play. I have personally experienced their penchant for curiosity a number of times as I have reported earlier (especially the past 2011 season with four deliberate changes of course on their part to swim directly to me and pass under my kayak at ten foot depth to get a good look at me) and I have also seen them at play, in competition, in joyfulness and angered. As we spend more time under the water with them and have the opportunity to see the other 90% of their lives I am quite sure we are in for some surprises.

There is also a vast record of interactions between humans and whales, though anecdotal (not having scientific verification), still indicating a species that is intelligent and feeling on a very high level. Peter J. Fromm's books, *"Whale Tales"*, are filled with stories of interactive experiences between humans and whales that strongly suggest higher intelligence...and desire to communicate.

I will mention one more thing in regard to what I have presented in support of the idea that the cetaceans are our planetary **OTHERS**, totally sentient, but living in a world so different than our own that their existence as the other intelligent species on our planet is unrecognized and essentially disregarded.

Communication and/or communion between our species on a delightfully but fictional level of actual "talking together" is remote at best, but is that all that communication/communion implies? Is it not communicating when one species approaches another out of curiosity? Is it not communication when an entangled whale is rescued by humans and when released, the whale approaches each diver and nudges them as if to say thank you, is it not communication when a whale lifts a diver in the air and gives them a gentle, albeit harrowing, ride on it's back? Is it not communication when a whale stays by the side of a boat in heavy fog and prevents it from crashing on the rocks, thereby saving the human on board? Is it not communication when I dive with the whales and send out loving energy as they pass by? I confess to these all being rhetorical questions. After over thirty years of on again off again interaction with the great whales, there is no question in my mind of their higher intelligence and gentle spirits.

My personal journey with "The Whale People" has convinced me of the need to have written this book. It is time for the human species to recognize that we are not alone in this Universe. We do in fact share the joys of knowing ourselves and our relationship to others with "The Cetacean Nation" and it is time to communicate that knowledge to them by respecting their existence as we do our own.

Chapter IX
Implications

Let us suppose now that you the reader are somewhat convinced, or at least intrigued, by the idea that the cetaceans are in fact an **OTHER** highly intelligent and sentient species upon our planet. I am certainly not alone in suggesting this as evidenced by the comments made by those professionals in the field who have been around these creatures and have come to be moved by them as I have. This has also been a favorite theme for a number of science fiction writers and the scenario for more than one film. Again, suppose we accept this emerging view of their higher intelligence (and I emphasize once again this higher intelligence is not the ability to balance balls on their noses, distinguish colors, numbers etc., I mean higher intelligence in recognizing self and all that implies!), then we are, as an intelligent species ourselves, obligated to evaluate more carefully our present treatment of them are we not? After all, is not intelligence and the enormous value we put on it as human beings, both as a quest for knowledge and a spiritual blessing, worth cherishing and honoring wherever we may find it? And if that is so, then we must look carefully at our present relationship with the cetaceans for we have come a long way from the days of The Delphic Wave when our ancestors respected and honored the cetaceans as "The Whale People".

For most all the world's people there has been a slow but hopeful reawakening of The Delphic Wave and this has manifested as a demand for the end of outright killing of these creatures for industrial products and food...or even science. Yet today arguments still take place within The World Whaling Commission as to whether whales should be taken (killed) in small numbers for scientific and/or limited food/industrial purposes. This is not however, though criminally reprehensible (if one is to accept that they are indeed intelligent), where the greatest danger to the cetaceans lies.

The migration of the great whales from feeding grounds to breeding grounds is an absolutely essential element in the life history and the survival of the great whales. There is no choice in this; they like ourselves, are limited in their survival by the constant need for acquiring food and reproducing young. This is what of course natural selection is all about, survival of the species. Each and every animal and plant species on the planet is held to that prime directive. For the great whales, the migration mentioned above is the strategy they have perfected to insure their survival within the limits of the resources available to them within the boundaries of their marine world. There are a great many factors that are at play in regard to these annual migrations; water temperature, salinity and a host of other factors including attacks by both sharks and Orcas. Yet the great whales have successfully sustained their numbers for literally millions of years by acting in harmony with these outside forces and fulfilling the demands of natural selection - they continue to reproduce and survive.

This is again a time to reflect on the enormous differences in a terrestrial versus a marine environment and the unique demands placed on a species within those environments. Whereas we terrestrial creatures that have evolved hands with opposable thumbs, thus giving us a significant control of our environment through technology, hands and fingers are not of any value when living in a free flowing liquid environment as the cetaceans do. Without this technology it is necessary to adapt to the conditions that exist, which is exactly what the cetaceans have done so successfully. Indeed so successfully that they have evolved higher intelligence during their evolutionary journey.

By necessity this yearly migration also takes place relatively close to shore, usually within the area of The Continental Shelf (that part of the earth's continents that is always under the surface of the oceans but generally does not go deeper than 600-900' before rapidly descending via The Continental Slope to The Abyssal Plain thousands of feet below). This is due to the simple fact that that is where (The Continental Shelf) marine food resources exist in greatest abundance. As stated in an earlier chapter it was the discovery of these migration routes that made the whaling industry so successful. Once the whalers knew where the whales would be in greatest abundance they could literally wait for them to arrive and kill them in large numbers most efficiently.

Since then scientific research has established the major migratory routes of a number of whale species and this has in some cases resulted in restrictions placed on human activity along those routes. Here in the north pacific where I do my work, The SPLASH Program (2004-2007) mentioned earlier, gave us definitive routes and numbers of humpback whales associated with those routes. It is now accepted that within the whole of the north pacific basin, there are only four viable breeding/calving grounds for the entire north pacific humpback population. That entire population also needs to access northern and arctic waters during the summer to engage in gorge feeding on krill. This gorge feeding is not just for the thrill of having an abundance of food, it is essential for the fattening of both calves and mothers, as well as pregnant females for the successful nurturing of their young as they begin their arduous life at sea.

It is absolutely clear then that there are only four breeding and calving grounds in the entire north pacific basin and these are, and have been, used for millennia because they provide the necessary elements for successful reproductive activity. I cannot over-emphasize the importance of this. Remember, they live in a different world, one devoid of technological methods by which to enhance reproductive potentials or utilize a myriad of locations (as we humans, through the technology of massive irrigation, can turn non-productive areas into veritable oases) as food sources. They must migrate between the feeding and calving grounds to sustain the species. And this is where a major problem arises. We humans use the same continental shelf area for our fishing, transportation, recreation and agricultural/industrial (in terms especially of consequences from runoff) activities - and there lies the problem.

Each one of these activities in it's own way, and with dramatic detrimental consequences to the whales, is becoming a more serious and immediate problem for "The Whale People". They have no way to appeal or fight back and this is exactly why John Lilly suggested the concept of "The Cetacean Nation" to actually be recognized by The United Nations as a singular intelligent entity and to come under it's (The UN) protectorate.

I will briefly describe the difficulties to the cetaceans that are specifically engendered by the human activities mentioned above.

The human fisheries industry has always been and continues to be fundamental to human survival on the planet. In order to keep up with the growing demand for more and more food resources from these fisheries (ostensibly because of increasing human population demands but actually due to insensitive and greed driven manipulation of the distribution of food on the planet in the first place), more efficient methods for gathering the fish have been developed (technology again). There are two of these fishing methods that directly affect The cetaceans and are increasingly detrimental to their survival capabilities. These are long lines (a line of hooks extending sometimes as far as five miles or more) and drift nets (fine mesh nets that hang from the surface of the ocean to as much as 40' below and also extend for miles). It is estimated that at the very least there are 1 million of these nets in the north pacific alone. The problem here for the whales, and everything else that swims in the oceans, is entanglement.

It is not difficult to imagine the panic of any animal but most especially an air breathing animal in water, when finding itself entangled in a long line or drift net. I clearly remember as part of my lifeguard training, being wrapped up in a tennis net and required to extricate myself while in the water. It was very frightening even with others standing by to help. Recall that sight is not the sense used by creatures of the sea, it is sound, so entanglements are not rare occurrences, and with the nets/lines placed in the same areas as migrating whales, entanglements are becoming more prevalent. In the eight years I have been working in my area I have witnessed two entanglements myself and heard of three additional ones in my area since 2007. Three fortunately resulted in rescues but two resulted in fatalities. As mentioned earlier, as food demands increase (though fabricated), so will the number of nets and long lines, and logically therefore, more entanglements.

A humpback whale entangled in a large net. I have personally come across three humpbacks in this situation. We all carry rescue gear with us now but rescues are dangerous and difficult.

More often than not a rescue is not feasible and eventually the whale dies. I came across this adolescent humpback dead on the beach just north of my village. It had gotten entangled in a long line and had been unable to move it's pectoral fins or it's flukes. This is very painful to see and is without doubt going to occur more as the demand for fish rises.

In regard to negative effects created by human transportation activities it is the simple fact that we double our boat traffic every ten years and this is becoming a problem for the cetaceans for two reasons, one: boat strikes (50% of the Humpbacks identified during The SPLASH Program had evidence of scars from boat strikes and/or entanglements) and the other, which is of even greater concern to Chris Clark at Cornell, is the increasing amount of ambient noise filling the oceans basins due to increased transportation noise and other factors that he refers to as "ocean smog" (the collective noises from shipping traffic, oil and gas exploration and production, and recreational traffic).

As examples of the first problem regarding human transportation affects on whales I recall my first on site observation of the damage that can be caused when in 2005 I came across a large adult blue whale that seemed to be logging (lying still on the surface), but on closer examination, it was clear that it had been struck by a boat's bow or been run over by a boat's propeller.

There was an open and very nasty gash just forward of it's peduncle area (the narrower area between the flukes and major part of the body) and I was quite sure from it's movement that it's back had been broken. Those of us that are engaged in photo ID work try to acquire flank shots (photos of the whales sides) as well as the identifying fluke shot so we can better determine the whale's health over time and determine damage by strikes.

Regarding the second issue around transportation is one that particularly concerns Chris Clark. As Clark has discovered, "many whales have very traditional feeding grounds and their migratory routes occur along shallow coastlines, which are now some of the noisiest, most heavily impacted habitats, but often it is along these routes that the male songs are sent long distance to prospective females, who might not receive the message through the "ocean smog". If females can no longer hear the singing males through the smog, they lose breeding opportunities and choices. The ocean area over which a whale can communicate and listen today has shriveled down to a small fraction of what it was less than a century ago. We now have the ability to fully evaluate where they are and how long they sing for," he said. "We now have evidence that they are communicating with each other over thousands of miles of ocean. Singing is part of their social system and community. Whales will aim directly at a seamount that is three hundred miles away, then once they reach it, change course and head to a new feature. It is as if they are slaloming from one geographic feature to the next. They must have acoustic memories analogous to our visual memories," he said.

His ultimate and disturbing conclusion (remembering that a whale's consciousness and sense of self is based on sound, not sight) is that, "we are slowly, inexorably, raising the tide of ambient noise so that their worlds are shrinking just to the point where they are dysfunctional". Clark's concern is not in the least exaggerated and is to be taken very seriously if we humans are to sincerely accept them as **The OTHER** sentient beings on this planet!

As with boat traffic in general, human recreational activities within migratory routes is also increasing and resulting in serious problems for the whales. Once again it is essential to remember that the migratory route is by necessity relatively close to shore and cannot simply be moved further out. More and more people are demanding closer personal experiences with the whales, whether on a whale watching boat, while fishing or just recreational boating in kayaks, on jet skis or other power craft. The result is that too many people are getting too close to the whales and that is having a negative affect on their behavior due to everything from simple annoyance to serious stress and even occasional physical harm (boat strikes etc). I have personally seen on some of my research runs as many as three boats and half a dozen jet skiers within a few tens of feet of the whales, even mothers with calves. There is an ad in one of the slick magazines in Cabo San Lucas showing a mother and calf with jet skiers literally inches away, suggesting the experience of a lifetime.

In 1998 when I was part of the crew of the research vessel *Dariabar* in The Hawaiian Islands monitoring humpback behavior during The ATOC experiment, we saw very clearly, on our sophisticated hydrophone images, and much to our surprise, that what annoyed the humpbacks most were jet skis!

One last activity that I will discuss briefly is industrial and agricultural runoff...pollution. Roger Payne of Ocean Alliance has completed a five-year expedition designed to establish the first baseline levels of synthetic pollutants in the ocean. Long-lived industrial pesticides, such as DDT and PCBs, re-concentrate as they move up the marine food chain and the whales are at the top of that chain. "Insect repellents and insecticides which have been spread on fields on land have now gotten out to whales in mid-ocean," said Dr Payne. After taking over 1,000 samples they found that all those so far tested showed high accumulations of the industrial chemicals, DDT and PCBs. PCB toxicity is defined as 50 parts of contaminant per million parts of animal, (50 milligrams per kilo) tests have revealed up to 400 ppm in killer whales, 3,200 in beluga whales and 6,800 in bottlenose dolphins. Contaminants such as PCBs and DDT have been shown to inhibit a mammal's immune system, its ability to function, and the development of its young.

While some whale populations are recovering since the 1986 moratorium on commercial whaling, anthropogenic (originating from human activities) influence may play a decisive role with populations that are at critical levels and endangered, such as the northern right whale.

In 2008 while searching for humpbacks in my grid I came across two whales that acted particularly lethargic. As I got closer to them I saw that both had a large number of badly infected skin lesions, some open wounds and a pasty, slimy overall appearance. Their backbones and ribs were also highly visible indicating poor nutrition. I ID'd them and then put them up on my computer to enhance the pictures I had taken. I realized on closer examination that they were gray whales. We rarely if ever get gray whales into the Sea of Cortez because they are bottom feeding whales and strain sand of crustaceans for their food. Since the Sea of Cortez is a deep and rather steep sided crack between The North Pacific and The North American Tectonic Plates there is little flat area for sand build up. Therefore, we don't see them here because there is simply no food.

Shortly after, I received an e-mail from my friend Dawn Pier in Cabo Pulmo, asking if I had seen any gray whales in my grid. She was as surprised as I for she had a group of them in her area and they too were looking very lethargic. They remained in The East Cape Area for a short while and then were gone. Later I learned that researchers in northern California had discovered the problem. For unknown reasons the northern California off shore sands were particularly badly contaminated by ag/ind runoff that year and the crustaceans had died...

therefore no food. The grays had come into The Sea of Cortez in the hopes of finding additional food since much of their feeding ground had become contaminated. This kind of thing is happening in many locations throughout the world and is becoming an increasingly serious problem.

The undeniable bottom line is once again the over population of the planet by we humans. The critical factor here is that "The Population Problem" is a monster with two heads. One is "People Overpopulation", created by too many people in a given area for the basic resources of that area to sustain the population in a healthy manner. These resources being clean water, sanitary shelters, sufficient food, available medical attention, and reasonable expectations for a healthy and rewarding life. This is primarily a "Third World Problem" but as we are all brothers and sisters on the same planet, it is our mutual responsibility to see these basic needs be made available to everyone. A big part of dealing with this is of course planned parenthood and birth control.

But this is not the only problem (people overpopulation). The other equally threatening head of the monster is "Consumption Overpopulation". This is primarily a First World Problem and stems from an insensitivity to the consequences of a consumer driven culture that does not recognize that it's demand for things is rapidly decreasing the sustainability of a healthy population within it's own ranks and certainly the rest of the world. It is this two headed monster that continues to wreak havoc in our present world and is at the core of our environmental as well as social problems. My ideas on this come from seventy years of active life and travel, twenty-three years of academic study and thirty-five years of teaching interspersed with field research. This is however not the purpose of this book and will have to be perhaps a future endeavor.

Suffice it to say then that The Cetacean Nation, though in some cases showing increased numbers and apparent good health (the conclusions derived from the three year SPLASH study indicated that at least for the north pacific humpbacks, their numbers were higher than expected) it is also very clear that if we humans continue to carry out our activities within the marine environment in the fashion we have thus far, the cetaceans will suffer accordingly. And here is the crux then! If the cetaceans are **The OTHERS** in the sense of a separately but equally evolved intelligence on this planet, then we humans have an obligation and moral duty to recognize them as such, indeed as bothers and sisters on this great journey to higher consciousness.

As we continue to find more extra-solar planets (planets outside of our solar system) in our galactic neighborhood, 854 at recent count, we are certain to find some with conditions similar to that of earth's. If they have liquid water, an atmosphere (preferably with about 20% oxygen) and mineral elements, it is very possible, indeed probable, that there will also be life forms on those planets. One need not envision "Men from Mars", simply life of a different order and perhaps...intelligent.

Chapter X

Discovery, Transformation and Healing

I have made my case for what I believe is the unique place that the whales occupy on our planet. This belief is based on what I consider reasonable evidence, both scientific and anecdotal. Now I would like to conclude with a more intimate and very personal description of what these unique animals have done to bring about a healing in me that was totally unexpected in my life at this time.

I will begin by repeating what I have said before. Though my journey has had it's difficult moments, with more most likely to come (but compared to the trials of literally hundreds of millions of people before me and after me they are trifles), I know this...like most people, my hardships tend to take on an exaggerated significance because, well...they are mine!

The journey that I have described in *The Others* is one of many elements. We humans are indeed a complex life form and this is manifested in all the different, colorful, painful, joyous and dramatic moments we create for ourselves during our individual journeys. I would like to address the three elements that I have found so dominant in my life and how my interaction with the whales has opened doorways for me to better understand my current time on this earth in the context of these elements.

Discovery

When I think of those discoveries that have significantly changed my perceptions over time and helped direct or redirect my life, I tend to see them not so much as a result of the educational process (which I honor and revere), but as one inspired by contemplation. As a child I remember being strongly influenced by those characters, either in books or movies, that were described as being contemplative - and I always found these characters to be my favorites.

As I made my way through the educational process I found, somewhat to my dismay, that academics held little interest for me and that contemplation, especially when in the outdoors, generated profound discoveries that far surpassed what I was being taught in the classroom. Of course I realize full well that a great deal of this lack of inspiration was due to my adolescence and indeed my drifting off into contemplation when I should have been more intent on the lessons. Nevertheless I was able to navigate the halls of academia efficiently enough to acquire the degrees I thought I needed to be a success in my chosen field of biology. Much of the inspiration for that, at times difficult path, were the memories of my discoveries on the banks of San Francisquito Creek as a nine year old, and the belief I maintained that someday I would rediscover this fascination with nature...and the critters through my chosen profession.

Since my focus in graduate school had been microbiology I had little opportunity to exercise my deeper interests since I was always in the laboratory and always looking at very, very, small critters. As stated earlier in the text, my interest level waned rather quickly and it was shortly thereafter that I found myself working at The UC Bodega Marine Laboratory as their collector. The discovery that I had while working there had a number of parts. One: I realized that I loved to work outdoors and this would be a necessary part of my life if I was to be fulfilled, two: that though harboring a deep seated fear of the ocean, I could overcome it, three: and most important to the theme of *The Others*, I had met a benign and intelligent critter that up until then I had ignored as being an academic relic of my university years.

It was not only that I had simply experienced the presence of a whale; it was the context in which it took place that led to one of the most important discoveries of my life at that time. I perceived in my deepest core that this gentle creature had intentionally guided my boat to safety and therefore must have certain cognitive abilities that at the very least were intriguing. The idea that this giant of the sea actually understood my predicament, evaluated the danger I was facing and then decided to help, penetrated my academic training like a bolt and opened my heart to interesting, if not fantastical, possibilities. Why it was that this discovery did not change the course of my life right then and direct me on an immediate and straight path to the study of the cetaceans, particularly the great whales, I cannot fathom. My suspicion is that I was not ready to embrace the significance of this unexpected discovery in a spiritually centered manner. I could do the biology, but was that what this had been all about? I do not mean spirituality in the context of anyone's religious dogma. I have never found religion to be spiritually, intellectually or emotionally satisfying. I mean it in the context of personal relationship to what The Huichol of Mexico (a native culture of the highlands) call The Great Mystery...all that is.

The discovery that we were perhaps not alone on this cosmic journey nevertheless had a profound influence on the direction that my intellectual and spiritual path was to take, and I became more aware of my part in the all of which The Huichol speak. A bit of personal research into the cutting edge of quantum mechanics and cosmology tended to support this developing premise that indeed, perhaps the entire Universe was "alive" with this essence (I refer specifically to Robert Tobin's book, *Space-Time and Beyond*). My realization was nothing new to philosophers, scientists, spiritualists, artists etc, but it was new to me in a very personal way and rewarding way.

This discovery became the seed for a more tolerant approach to things fantastical, including the idea that we humans were not the only sentient species on the planet. This is has been a favorite theme for science fiction writers for decades and is a major tenet of *The Others*. Though the experience, and my discovery were dramatic enough by themselves, there influence in my life for many years remained subtle indeed. I became more contemplative, more curious and certainly more enthralled with life in general. At the same time however, the reality that my life was becoming more restricted and now included the responsibilities inherent in adulthood and parenthood, offered me little time to engage fully in researching the possibilities suggested by my discovery.

My discovery did nevertheless prepare me for my future in the classroom in that I always tried to tie in the scientific facts with logic, reason, contemplation and spirituality. As with my feelings about organized religion, I feel the same way about spirituality. There are hundreds of ways that it can be seen and exercised. No one, no group has the ultimate answer to The Great Mystery...at least for now in my opinion. By now the extent of my hubris must be literally, no pun intended, and I hope not painfully, running off the graph. Nevertheless, I was told years ago by ChoQosh Auh'Ho'Oh, a Seneca/Chumash Medicine Woman who attended my classes and became a cherished friend, that it was, "time to speak my truth". I have tried to do so ever since.

Transformation

Though a transformation of my identity was subtly in progress after my 1971 whale experience, it would be some time before an actual, acknowledged change would become clear in my mind and would compel me to manifest a conscious redirection of my entire life path. Indeed, I would not see another whale for twenty years (1992) and it would be thirteen more years (2004) after that until I began my intimate connection with them on a regular basis and the healing would begin.

Transformations can be chaotic; three marriages, two children, three stepchildren, various and asundary jobs, college teaching, single parenthood, illness etc all had their means of testing anything I may have learned from my experience with the whale in Bodega. In a sense I was becoming emotionally and physically numbed. I was following a routine that worked but only if I didn't ask myself, "Urmas, is this it, is this how it's all going to play out"? I was pretty clear in regard to the fortunate reality in which I lived. I was comfortable, enjoyed my teaching, and I easily embraced new information that capitalized on my whale experience but nonetheless, I was unsettled and finding it difficult to turn my newly gained understanding into a viable and powerful existence.

Fortunately I had a great deal of help along the way. I have mentioned a number of people and events that aided my transformation and without them it would all have been much harder. In 1986 I was asked to teach a course in Appropriate Technology at the college extension on the Indian Reservation in Covelo in Mendocino County. As a result of that assignment I became aware and eventually familiar with more of the history of "The First People" and the individuals themselves. Once again I found myself opening up to very new orders of mystical experience and always in my mind was the discovery of the basic tenet that within this very real universe was an entire other universe that I was just beginning to get a glimpse of. Through a number of different Native American teachers I found myself moving more and more into a realm that had held interest for me but eluded me when I had been younger. This realm was the inexplicable, the fantastical, the spiritual and now the seemingly more attainable.

There subsequently came another series of personal setbacks including another divorce (my third), the death of my mother, whom I had been very close to, and my doctor's warning that I should have my prostate biopsied because of cancerous indications. I was a bit staggered by the combination of these events and often found myself often depressed. I continued to teach and in a very real sense it was my salvation for a time. I personally felt that I was reaching an internal peak of effectiveness in my lecturing and I also knew that soon I would have to leave the profession in order to retire when I felt at my professional best. In a sense it was as much as anything else, a recognition that I desperately did not want to continue teaching after "losing my edge".

By 2000 my third wife and I had purchased some land in Mexico and had hauled an old trailer there. Now that I was alone however, I wondered if it might be possible for me to find the peace and empowerment I so needed by living six months a year in Mexico and six months in northern California. I was able to make that happen in 2004 when I came down to El Cardonal with dreams of writing, SCUBA diving, kayaking and just being in the tranquility of that small Mexican village. Slowly, as I began adjusting to the pace of life and the constant and intimate contact with the sea, the desert and the mountains, I began my transformation. I literally started becoming another person. Years of protective psychic armor fell off and a different Urmas began to appear. Five years later I made the decision to live in the trailer in Mexico full time and I cut all ties to The United States and devote my energies to the whales and the other critters that I was getting to know.

I was now sixty-eight, and here at last was my transformation. I had utterly changed my physical reality and reinvented my world. Not in just moving to Mexico but in deciding to attempt to self-heal the problems that I mentioned earlier (prostate, shingles, kidneys, skin). To a very real extent I was guided towards this transformation by reading the works of Louise Hay and others. I knew it was going to be difficult, perhaps impossible, and that there were no guarantees that I would succeed, but I talked it over with my kids, explained my reasoning and much to my joy, they agreed to "let me go" on more levels than just moving to Mexico. I had already been strongly moving in this direction in 2004 by the end of my first six-month stay in Mexico. In my life style in the village of El Cardonal I experienced the peace and feeling of belonging that I had always known was possible but until that time had been difficult to attain. My pangero Vicente started calling me Herme (he still can't get his tongue around pronouncing Urmas) and his little daughter gave me the nickname "El Ballenero", The Whale Man. For all intents and purposes I had found "Shangri-La".

Healing

I have already described the evolution of my scientific interaction with the whales as a chronology of events and the eventual success of the work in helping initiate a MMPA for this part of Mexico. I have also addressed at length my evolving belief in the cognizance of these creatures and the need to develop a more realistic and respectful attitude towards them as **The Other** sentient beings on the Earth. I will finish by trying to articulate the healing I mention in the sub-title. As a seventy year old I have lived and experienced enough life to have a history. Mine isn't really all that different than most folks-to some extent colorful, certainly in many ways rewarding and satisfying. My greatest accomplishment has been the relationship that I have with my daughter Kersti and my son Zack who are both healthy, intelligent and good souls.

Nonetheless, my transformation also made me vulnerable to the bittersweet pain that comes from dropping one's psychic armor and seeing that the armor had been only partially effective...there were scars. Some of those scars were relatively recent and deep. Now came the healing. I was pretty confident that living here in Mexico, transforming my reality and delving more deeply into my initial discovery of the infinite possibilities within one's reality (via my experience with the whale in Bodega years before), I would find the peace I had been seeking all along. I even made a sign at the entrances to my sanctuary where it welcomes all who come in peace- "Shangri-La", from the book by James Hilton, *Lost Horizon*.

It was not however until I began to swim with the whales that I really began to heal. Being near them in a panga was incredible, in a kayak even more so, but to be in the water with them, in their world was extraordinary. I will try to describe what it feels like because I know for a fact that being in the presence of "The Whale People" in their world has gone a long, long way to help heal the boy whose life started with the sounds of bombs and the palpable fear of all those around him...my inner child. All the fears, failures and insecurities that I carried with me all these years had to be released if I was going to join my new friends in their world. The first time I dove with a whale, my dear friend Susan Janssen took a picture of me just before I went in. I asked her later, "what do you think was going through my mind at this time" - her answer was immediate and so insightful, "Am I worthy". Every time I have gone in with the whales or am in their presence I ask myself that question because Susan was right on, that's exactly what I feel. Making the decision then to go in is an affirmation on my part that yes, I am worthy, and no, I have not conducted my life perfectly, but I ask that you accept me as I am and mentor me to become what I hope to be...whole again.

The physical feeling I have when I am actually in the water with the whales is of itself healing. Their energy can be felt, and in my reality that energy manifests as curiosity, ancient wisdom and love. I know that many of my colleagues, known and unknown, scoff at such ideas. That matters little; indeed it matters to me not at all. When I have swum with mother and calf humpbacks, cruising blue whales, Mobula hunting Orcas or playful sperm whales (see appendix A for full stories) I feel surrounded by such gentle strength, such loving acceptance and such spiritual awareness that I cannot but be affected. Are they aware of the healing that they are extending to me? I don't know, perhaps one day I will find out.

There is an another feeling that I would like to mention. When in a panga near the whales (I only do this when I am actively seeking a photo-ID), I feel a physical thrumming, a deep bass vibration in my body. Subtle of course, but definitely there, and it is coming from the whale. When I encounter a whale in my kayak the thrumming increases in strength and I feel as though it creates a bubble of silence, peace and communion, and that too is healing. I have been in that state before through deep meditation and also upon the occurrence of certain events in my life that defied simple explanations. One can imagine then how I feel when I actually enter the water with them, dive with them and swim along side them with their bodies a mere 10' away. Everything but that moment becomes unreal. It is as though the bubble I mentioned a moment ago becomes the entire universe and I am privileged to experience the power of The Great Mystery...and that is healing!

I will end here then. There is no attempt on my part to suggest that I have achieved some extraordinary connection with 'The Whale People" or that I have found enlightenment or "God". What is important to me is my belief that **The Others** have opened their hearts to me and in their welcoming they have helped me to heal. I simply wish to share that joy.

I would also like to share a few of my favorite pictures with you...enjoy and "que les vaya bien, mis amigos todos"!

This is one of my favorite pictures. It is a male humpback in full breach. He came from nowhere and we stayed with him long enough for an ID and for me to take a little swim with him a bit later.

This is a wonderful and valuable "flank shot" of a humpback mother and calf. Though the mother would not fluke up, so a definitive photo ID could not be taken, the flank markings are nevertheless clear and allow us to identify these two whales as related if seen together again.

This is a good example of a humpback preparing to do a "tail lob". In the southern breeding grounds this is generally, but not exclusively, done by males to announce their presence. It is also known to be a means to panic and school fish for consumption. This whale tail lobbed eighteen times in succession. A good half to two thirds of their body is out of the water and the sound can be heard even through the air for miles.

Here I am diving with a pod of Orcas as they hunted Mobulas. This was a singularly delightful experience for me. Imagine the scene in the movie *Never Cry Wolf* except that it is Mobulas and Orcas not Caribou and Wolves. See story in appendix A.

This is mature blue whale cruising south in The Sea of Cortez. This is the largest animal that has ever evolved on our planet. They may reach a length of 85-90 feet. The record is 110 feet. Among some of my Native American friends in Northern California they are referred to as "The Wisdom Keepers". I had the distinct honor and pleasure of swimming with this one three years ago.

I am sitting on the bow of Vicente's panga and being treated to a full display of surface behavior when this whale breached once more and seemed to wave to me. After obtaining a good ID shot, I dove in with him and we swam together for a short while. After getting his picture on my computer I recognized him as "Odin" the whale my son adopted in 2008 and that has returned to my grid for four different seasons.

Appendix A
Unforgettable Experiences

Before I relate a number of stories that will help to understand my personal journey with "The Whale People", I would like to tell three stories that each in their own way molded the personal/spiritual perceptions that would help set the tone for my adult life.

Premonition of Death, 1963

After High School graduation in 1960 I began working summers between University years as an apprentice lineman for Pacific Gas and Electric Company (PG&E). It was a job found for me by my then future father-in-law whose motives were unclear. The work was minimally dangerous (I was on a crew whose task it was to clean and "change out" the insulators holding 110 and 220 KV electric lines to the 60'-100' transmission line towers throughout The San Francisco Bay Area) but paid very well for an 18 year old in 1960...$22.00/day!! Though my mother was less than happy with the danger involved, her father, The Sea Captain, told her that when he was my age he was climbing the rigging in tall ships going around The Horn. Fortunately she listened to him so there was never any need for me to rebel against her.

After three summers of working for PG&E a number of us had developed faster and faster ways of climbing and descending the towers. This was unnecessarily risky but we were young men full of strength, energy and a substantial dose of stupidity. One morning I was working alone on a tower and had my back to the towers leading to mine. Suddenly it seemed that all sound disappeared and I felt like I existed inside a bubble of utter quiet and timelessness. Then a voice inside my head said, "Woody (one of the crew) is dead"! I went into a semi-daze and then a few minutes later I saw an ambulance racing down the highway and in the back I saw Woody's face looking at me. This was of course impossible since the ambulance was a good distance away and had a very small window in back. At this point I was certain that he had fallen and was in fact dead. The feeling was absolutely clear and without qualification. I descended the tower and waited for the crew truck to arrive. When it did, two of my friends jumped out and before they could say anything I said, "Woody is dead isn't he?" They looked aback and replied that, yes he had fallen, and was killed instantly just a short time ago. Our foreman said it would be OK to quit for the day but we all agreed it was better to keep climbing. What that day taught me was that there were different levels of "knowing" and I had just experienced an alternative one to that which I had so far based my life on.

The White Wolf, 1990

One of my favorite spots in all the world is the high country of The Sierra Nevada Mountains of California. In 1990 I took my wife Susan up to The Sonora Pass area to do some hiking and fishing. We found a nice spot to camp and set up our tent. While gazing out at the meadow before us, I saw about 150' away, a white wolf. He was crossing the meadow from our right and heading towards a head high boulder to our left. Susan was looking the other way but I watched it for the few seconds it was visible and then I said to Susan, "damn, I just saw a white wolf out there in the meadow". I qualified that by saying that that could not be, because the area no longer had wolves; they had long ago been hunted out of existence by humans. The fact however, as I told her, is that I was certain it was a wolf.

I told her I was going to approach the boulder from the right and for her to watch to the left if I flushed it out (the boulder stood pretty much alone in the meadow). I was very hopeful that I would see it again and that Susan would enjoy seeing it and secondarily corroborate my story. No wolf, no nothing. No prints, no sign of any kind that anything had been there. I was nevertheless certain of my experience; I had been "connected" to nature since a small child and not about to mistake a domestic dog for a wolf. If there had indeed been nothing there, what an amazing hallucination!

The image stuck with me and just a week later the college gave me an assignment to organize an "ElderHostel" class. I decided to do one on The Native People of Mendocino County. The program was approved by the college and The ElderHostel organization and I began recruiting instructors. I wanted very much to use Native People as much as possible so I began asking my Native American contacts to recommend someone.

Yep, everyone said the same thing, "get a hold of White Wolf." Synchronicity? I did, I hired him, we became friends and after the class he brought me the body of a Peregrine Falcon that had crashed into his car window, conducted a healing ceremony and gave me the critter. I buried it and planted a Japanese Maple Tree over it. The last that I saw that tree, it was huge. White Wolf went on to receive a Master's Degree in Museum Administration and is now in Washington DC, and I believe involved in Native American exhibitions at The Smithsonian.

A walk with Tomas, 1995

Thomas Pinkson, a man I met at an ITA conference in Killarney, Ireland, had successfully completed a thirteen-year initiation into The Deer Clan of The Huichol Indians of Mexico. I invited him to join one of our expeditions to The Channel Islands in search of blue whales. In the main text you will find another story regarding "Tomas" but I would like to add this one, for it is also important.

One day while taking a break from the search for whales, we anchored off of Santa Cruz Island and Tomas asked me if I would like to accompany him on a hike into the interior of the island. I answered of course, and we had the mate drop us off on the beach to be picked up later. I told Tomas that according to my Native American friends the ancient Chumash Indians of Santa Barbara had buried their dead on that island. As we began our walk we passed through a small forest of short eucalyptus trees and were startled by hundreds of crows that exploded from the trees and took off for the interior. Tomas smiled and said,

"The Crow People are letting the spirits know that we are coming". As we approached the narrow entrance to the canyon that wound deeper into the interior of the island (there were two large boulders, like sentinels, just before the entrance) Tomas signaled a stop. He carried out a short water and tobacco ceremony to honor the spirits of the dead and then we proceeded.

I led the way and about a half mile into the canyon I spied a rock outcrop that seemed to beckon me. I had been a great fan of the Castaneda's books and totally understood the idea of a "Power Spot". I turned to tell Tomas that I was going to spend some time there and...he was gone! I have no idea when he left me or to where he went but I felt fine with that so there was complete accord in my spirit.

When I got to the outcrop I sat for a while, then peeled off my clothes and began a chant. For reasons I did not understand, I chose to remember that where I was born, Estonia, and the particular island my ancestors came from, Saarema, their totem was the bear, or "Karu" in The Estonian Language (a language isolate like Basque and Yuki). I began chanting Karu at the top of my lungs and then began to dance in a circle around a small boulder. The more I chanted and danced the more surreal the moment became. I was on an island, I could see the sea in front of me and my boat was anchored nearby...I was spiritually "home".

On the way back down the canyon I wondered where I would meet up again with Tomas and then I heard an eagle call, loud and shrill from above me. When I scanned the canyon rim, there he was...it was Tomas. Later when we met up, words were not necessary. From that time on I began relating to all the critters as "The "People". I knew from then on that the whales would be "The Whale People".

My very first experiences with humpback and blue whales during 1995-6 are recorded in the main text (chapter IV).

Hawaii 1998 ATOC Expedition and the first humpback encounter in my kayak

After the excitement of the passage from Mexican waters to Hawai'i we took some R & R on the Big Island. I have to confess to a feeling of pride and accomplishment, particularly when asked by anyone where we had come from. It was that very special feeling that I have known only rarely in my life. The feeling of having accomplished something special and learned things about myself and my reality that made me stronger. I truly felt different, tougher, more confident and happy with myself. I hadn't felt that way in a great long time! It was as though a fresh breeze of space and time had blown out some very old cobwebs.

During our time in the port of Hilo, I twice went out in my kayak to see if I could draw in a whale or two. One of my fantasies was to "look into the eye of the whale" (as the window to the soul), and at kayak level that just might occur. I had been eye to eye with humpbacks before, on the deck of *Dariabar* from a distance of maybe 4', but I wanted to really be eye to eye. I paddled out of the harbor entrance and into deep water. As I glided across a 4' wind swell I kept repeating the phrase, "I am here now, let us meet". I could hear their blows from various directions and the hard slap of their bodies as they breached nearby, but the swell hid them from view.

As the wind freshened and the swells got larger I started back in. One lone humpback began a parallel course with mine. For about 10 minutes we angled towards each other and then when he was only 25-30 yds away a fishing boat cut between us and he was gone. Above all else they hate the sound created by the cavitation of small outboard engines. Returning to the harbor I was treated to the company of a rather large leather back sea turtle. I was able to get eyeball to eyeball with him and I saw such an ancient wisdom looking back at me that I felt I was in the presence of a Shaman. Lord knows what he thought of me?

After a week in Hilo we set sail for Oahu and the port of Honolulu. We sailed out in late afternoon so we could catch the volcanic display of Kileau at night. I was woken for my watch and saw the fiery spirit of Pele as we rounded the south shore of Hawai'i. It was like a dream, how could I be so lucky as to be experiencing all these things and the voyage had only begun? We continued to sail along the west coast of the Big Island working our way north to Kiholo Bay in preparation for crossing the infamous Alenuihaha Channel to Maui'i. Just a ways out of the Port of Hawaihie we dropped sail and deployed our hydrophone unit to see if we could detect some whale sound. What we got was the most elaborate symphony of cetacean vocalizations I had ever heard. Danny piped the sound through our speakers on deck and for the next half hour or so we were treated to a most wonderful performance. Not a whale or dolphin in sight, the ocean dead calm, oh my, oh my. To envision the ocean below and around us filled with invisible singers has remained with me as a signature of the marine world.

We crossed the channel that had twice destroyed the invasion army of Kamahamaha I at night with good wind but mild seas. Next day we sailed past Maui'i and Molikai'i, crossed the Kaiwi channel and got to our berth in Honolulu in time to attend a super bowl party at the Waikiki Yacht Club as guests of David Lyman, son of Admiral Lyman of WWII Pacific fame. Our crew did itself justice in the general revelry as we were to some degree celebrities having made the passage in an El Nino' year with contrary weather. After a week in Honolulu making connections with Cornell, Greenpeace and the Hawai'i maritime museum we were ready to sail to Kaui'i, our research destination. My wife Susan Pepperwood joined me in Honolulu a couple of days before departure and sailed with us at night across the Channel. She joined me on my midnight watch and got a first hand experience of night sailing in the Pacific. Next morning we arrived at the harbor of Nawiliwili just outside of Lihue and tied up at what would be homeport for the next four months. After a short vacation with Susan on Kaui'i the real work was about to begin as Cornell Universities people began to arrive with their very sophisticated electronic gear.

On Tuesday the 17th of February I returned to the boat after a short vacation with Susan on Kaua'i. The crew and the contingent from Cornell, along with their gear, were busy in preparation for our first session on station. Our goal was to monitor the surface behavior and the vocalizations of the humpback whales to insure that there were no harmful effects from the experiment during the The Scripps Institute of Oceanography's global warming studies. There had been concern regarding the "big sound" and it's influence on the natural behavior of these most wonderful creatures. We were to be on station directly over the big sound source off the north shore of Kaua'i by 7:00 AM every morning for seven to ten day periods. This was to last until the middle of May at which time the whales would begin their great migration to Alaskan waters and the experiment would be terminated. Our equipment consisted of some of the most sophisticated listening devices available. This included a twenty seven unit hydrophone array that we would tow behind, and a number of sona buoys (pop ups, not sound generating) that we would deploy at strategic locations around the sound source. The pop ups would be anchored to the bottom for periods of up to three weeks after which they would be retrieved and their recordings analyzed in relation to the times when the big sound was generated. At no time did we on Dariabar know when the source was operating so that no bias of observation would be possible.

The Cornell contingent consisted of the project chief, Dr. Adam Frankel, and three to four graduate assistants for surface observation. I have never worked with a more capable, cooperative and down right fun bunch as these young people. We left Nawiliwili harbor at 7:00 PM that evening in order to be on station the next morning. My first watch in some time was pleasant except that I had already lost my sea legs and I was a bit woozy for the first few days. We hove to over the station at night and I was treated to very calm conditions on my watch and was able to simply lie on deck and appreciate the starry silence. Carl Jung's concept of the sea being the window to the subconscious was strongly reinforced during these periods of solitude, broken only by intermittent and always stimulating conversations with my helmsmate Jack Frost. The roll of the boat, the smell of fresh sea air, endless stars and the occasional sounds of marine creatures were truly ethereal. On Saturday the 21st the weather freshened and our work became unproductive so we sailed around the west side of the island and headed back to Nawiliwili. This allowed us a view of the entire Napali Coast from about two miles out and in moderate seas with the coast looking like something out of a fantasy. Incredibly beautiful, but lots of rolling sea and the discomfort that that implied for me. When we rounded the south end of the island we picked up the wind again and some squall activity so my watch at midnight was demanding. Nevertheless, Jack and I had a deep philosophical discussion that passed the time well.

We stayed in port for the next seven days waiting for the sona buoys to arrive and catching up on ships maintenance. During these lulls I found myself having the most vivid and extraordinary dreams. I was even gifted with two lucid dreams (which I had not had for some time). I, like the rest of the crew, also read voraciously, easily finishing 10 books during this part of the voyage. At this point in the voyage I began to accept that the reality of this experience was to be quite different than my original fantasies. My intention to find spiritual communion with the whales, and myself was taking a very different path than I had imagined. Simply wanting to be a good sailor, or wanting to convey to the whales my soul intent was not enough to make it happen. Yes...I was going to have to work for it - imagine that! In my journal the night before (Feb 28th) our second trip on station I wrote, "the wheels of greater understanding are clearly in motion". The second trip had many difficulties. The weather was excellent but the sophisticated equipment did not perform well so there was lots of pretty frantic activity for a few days; including a run back to port, loss of a pop up, repair of the array, but everyone kept it together and we still got in some real quality observation. Returned to Nawiliwili on the 7th of March feeling good in all ways.

During the break until the next trip out I discovered the amenities of the Marriot Hotel just a block from our berth. Presenting myself as a member of the elite clientele there I partook of their swimming pool and beach. Gaudy and very tourist oriented but a great place to hang out when not on duty. On the full moon night of the 12th of March my helmsmate Jack treated me to a moonlight performance on his pan flute. It was just the two of us on board and while I was enjoying the moonrise he had climbed to the top of the main mast and began to play. I cannot describe the beauty of the moment, only to say that it will remain as one of the purest moments of contentment I have ever known! We went out again on the evening of the 14th and arrived on station at the end of my watch. Dr. Chris Clark, the director of Cornell's bioacoustics lab and world's expert in cetacean vocalization, joined us on the 19th to help in retrieving all the pop ups we put out on the last trip. We got most of them but lost one. This trip out turned out to be even more eventful then the previous one, requiring another quick run back to Nawiliwili, loss of our RIB (rigid inflatable boat) in the surf in Hanalei and returning to station in a near gale. It was however some of the most exciting time I had at the helm. Surfing 20' swells was an experience! At one point I lost my concentration and the wheel spun out of my grip. I fell backwards and it could have been real ugly but I was fortunate enough to grab a spoke and it pulled me back upright. When we finally returned to port everyone was truly exhausted but the work had been done.

I took two days off to meander around the island, even rented a car. For some reason I just could not get picked up hitch hiking. No one else had that problem, really made me wonder? It was great to be completely on my own and I took advantage of the time. Even caught a movie in town. After one more very short, but incredibly intense sail out to station (primarily deploying sona buoys for a long term vocalization record) we again returned to port for a long R & R. This concluded the major portion of our work with Cornell so I took a long leave of absence to come home and revitalize my other reality. Upon my returning to Hawaii again in early June we retrieved the pop ups and sailed to Honolulu where I was to meet a group of students from Mendocino College for a ten day course in whale and dolphin ecology - and that's another story (see main text).

First close encounter with a gray whale, 2004

My second visit to Bahia Magdelena in 2004 wasn't so much to see the gray whales again but to visit the daughter of the hotel owner in San Carlos, the small fishing town on the east side of the bay. She was a delightful woman, pretty, well educated and old enough so I didn't feel inappropriate in my attentions. Her name was Maria (what else, eh?) and though nothing came of it we enjoyed each other's company for a short time. I did however decide to go out again to in fact see the gray whales. The first time had been a bit early in the season and I thought this time my experience would be more dramatic (after all, I hadn't "petted" the whales on the first trip). I got more than I asked for due to the fact that my pangero this time was a novice and he inadvertently took us right over the top of two males and a female engaged in mating behavior. The whales were not interested in us nor paying attention and so all three came up underneath the panga lifting us out of the water about 2' and then swam away. I have no doubt they were insulted by our rude behavior but they let it pass. I had been standing at the bow looking down so saw the whole thing develop and must admit that for a moment I was indeed concerned. My pangero looked sheepish and I reminded him about the laws governing close encounters. No one was hurt so no harm done but it was a sufficiently dramatic encounter for me to remember vividly in all my later encounters. We don't often get gray whales on the east cape, so any further experiences with them have been minimal.

I see my first humpback in front of El Cardonal, 2004

As I recorded in my journal that day, "major event of the day.... I saw my first whale here in our bay." I was out pruning and heard a whapping sound. At first I thought it was the guys next door working on clearing the property. Then I saw there was no one there. I looked out to sea and caught sight of a large white flash and again heard the whapping sound about where I generally kayak to. I figured it was a manta ray but one really big one. I looked more closely...it was a big whale breaching and twisting and the whap was incredibly loud from how far away he/she was. I couldn't get the binocs out quick enough except for one final glimpse as it sounded. It was really exciting to know that there were humpbacks right here in front of El Cardonal. Little did I know what this was to mean to me.

My first close encounter with a mother and calf humpback, 2005

2005 was my first real season of photo ID'ing the humpbacks in my area. My intention had been to assist Richard Sears by monitoring the blue whales in the east cape region between Punta Perico in the north and Punta Arena in the south. With me on those very early searches was one of my interns from California, Lenee Goselin and her cousin Kristine from Australia. On one of those searches we came upon a mother and calf humpback in front of Punta Pescadero (a smaller point in the bay) and we were all delighted and moved by the sight. I asked Vicente, my pangero, to parallel the whales so I could get a flank shot of the two together. A flank shot is meant to ID two whales traveling together, particularly a mother and young calf for future studies of migratory and/or long-term relationships. A mother generally does not dive (fluke up) when she is with a young calf for fear of leaving it on the surface for very long because the calf doesn't know how to dive deep yet, and is vulnerable at the surface. Due to this it is often not possible to get a good fluke shot on a mother during this time. The fluke shot being a photo of the underneath of the flukes (tail) where each humpback is distinctly marked.

After I got my shots of both flanks I was surprised to see the mother change course and swim towards us with her calf. Gray whales are known for this but not so with humpbacks, they are far more reticent to make contact with humans. I asked Vicente to shut down the motor and we drifted as the whales approached. The mother circled the boat with the calf nearby and then stayed about 50' off as the calf came closer and examined us more closely. So closely indeed that the calf bumped the gunwale of the panga, turned it's eye to us and treated us to a fine spray of "whale breath"! Exhilarating experience, albeit a bit rancid.

Blues and more blues, 2005

The daughter of a good friend, Kirin Riddell, who had just begun studies in Environmental Biology at Stanford, was visiting her parents at their home across from me in El Cardonal. I really enjoy giving people "The Whale Experience" (being up close to whales) and since Kirin was a student of biology I asked if she would join me one day for a search. She was more than willing and we began the day with a humpback encounter within fifteen minutes of being out. I was able to obtain a good fluke shot and then asked if she would like to go further out and perhaps find a blue whale. Need I ask, right? It was rhetorical anyway. She nodded enthusiastically and since the sea was utterly flat we took off for deeper waters (I had found by this time that if I were to ID blue whales it would require an expenditure of much more time and since "time is money" I rarely went into deeper waters).

When we were about eight miles out Vicente, he of the eagle eye, spotted some blows. They were yet a distance away but it was clear that there were more than one and certainly worth investigating. We hustled out to them and there they were, a group of eleven blue whales of all sizes, ages and genders, leisurely cruising south. I was able to obtain good ID's on all eleven and we had a memorable afternoon to say the least.

I asked Vicente to get us in the midst of them so I could shoot to both sides. Soon we were surrounded by blue whales on both sides, behind and in front of us. Some came real close to the panga and as we worked our way through the group I felt elated in a way I never before felt. These were the biggest creatures the earth has ever seen, and I we were in their midst. I did not start diving with whales until two years later so I missed an incredible opportunity, but whose to complain, eh?

To top it off, on the way in we came across two more humpbacks and I was able to get fluke shots on both… what a day!

My first encounter with sperm whales, 2007

Late in the 2007 season, April 6th, Vicente and I were searching off Punta Pescadero for humpbacks and I decided to go into deeper waters to see if any blue whales were passing through. When we got out about ten miles Vicente spotted some blows but they looked odd to me. As we approached, Vicente said "Cachalotes" (sperm whales). I had never seen a sperm whale (we're talking Moby Dick now, their blow is very different due to only one blow hole located slightly to the side of the head so the blow is diagonal) and they are toothed. There were four of them "logging" on the surface after a deep dive and getting their breath back. We were in about 1500' of water and they had probably been deep diving for humbolt squid in the area. My friend Nadia Cisneros, who was working on her PhD at Scripps, focused on sperm whales so I thought I would get her some ID's. The whales were cooperative, one even turned and came for us (talk about having images of Moby Dick) but only out of curiosity. I got my ID's and we moved on. Later Vicente asked me why I didn't dive with them (I had begun diving with whales that season). Well...I tried to rationalize it all but the truth was that I had considered the fact that they were toothed whales (like Orcas) and could really do a number on me if they so desired. Vicente understood but I vowed I would not miss the chance again. It wasn't until 2011 that I got the chance and this time I took it for the "Dive of a Lifetime".

"Perseverance Furthers", kayaking with a mother and calf, 2007

By late March of 2007, I had acquired a number of good ID's and my research funds were pretty much expended so I spent more time in my kayak out at The Whale Zone (WZ, two miles from shore) where I had made most of my encounters. The winds had died and I decided to go out in the afternoon, which I rarely do since the winds generally come up then and paddling is not pleasant. At about one mile out I stopped to listen for blows. I had discovered that often one hears a blow rather than sees a whale since hearing is a 360 degree sense and seeing is around 90 degrees. A whale blow can be heard from a good distance, on a still day, perhaps as much as a half mile or more away. I didn't hear anything but my eye caught a large splash in the sea north of me around San Isidro (three miles distant). Then I saw another and a momentary large black object break the water...breaching humpbacks, no doubt! I paddled to where I thought was a good intersect point, having watched for a while and seen that they were moving south toward me and waited. There was a rare but occasional fog bank just a bit further out from me that day and I stayed on the edge of it hoping the whales would not venture into it and I would lose them. After waiting some time, perhaps twenty minutes it was clear I was in the right place and I could now see the blows and characteristic dark humped backs of three whales now only a few hundred feet from me. As I continued positioning myself I found myself closer and closer to the fog bank and as they came within 100', yes, they disappeared into the mist. I followed for a bit but lost them. I could still hear their blows and I knew they were close but I also knew that it was best I not go too deep into the fog.

I paddled back out of the fog, using the sun as a marker, and rested from my attempt for an encounter (by this time I had come to think of a "close" encounter in my kayak as being less than 20'). As I rested and enjoyed the feeling of having been in the fog and just hearing the whales nearby (reminiscent of my experience 35 years before in Bodega), I glanced up and damn, once again just a bit north of me two more humpbacks coming my way. This was great and a very rare occurrence in such a short time. I positioned myself again, paddling in a fury, and once again the whales disappeared into the fog. I had been close, maybe 50 feet or so but not a close encounter. I was a bit tired now so decided to paddle in, I mean how many whales could I encounter in a day. Well, here comes another one and I paddle like the devil to get to it but alas I am once again defeated by the fog. Now I am really tired! This time I am very ready to head for home because I have been paddling for some time now and still have a good distance back to my pull out. But no, here come yet two more whales and is there a choice? Of course not, I need to make contact. This time the two whales do not enter the fog (the fog actually had dissipated quite a bit) and I am in great position. It turns out to be a mother and calf. As I they come along side I paddle to stay with them and edge closer and closer to them. Soon I find myself only 10' or so from the calf with the mother (a very big whale) on the other side. I realize I am being completely trusted by the mother because the calf is very young and cannot stay down long. As we continue together for three blow series I feel that I have been given a blessing from "The Whale People".

I was so tired after this that I made it only half way to my usual pull out and beached instead in front of El Cardonal's one motel. As I pulled my kayak out, the Quebec family (later to become my "adopted family", The Q's) that stayed at the motel each year during winter came down to meet me. They laughingly told me how they had watched "the great chase" and how they wondered at my stamina; I told them it was all adrenalin, eh? The young daughter of my Pangero Vicente had also been watching this from her house on the hill and had run to her father calling, "El Ballenero, El Ballenero". I am still known by that name here in the village today, El Ballenero, "The Whale Man".

In the midst of a "competitive group", 2007

One beautiful day in the middle of March, Vicente and I were at the southern end of our grid and I spotted some blows a bit further south. Vicente confirmed them and gave me a thumbs up for my spotting (always a thing of pride for me when out with my piloto "Eagle Eye Vicente"). They were quite a distance away and it seemed there might be more than the common two or three (humpbacks are not often found in large family groups and it is thought by some researchers that it is a result of the massive destruction of their family units by whaling activities in during the 19th and 20th centuries). As we closed on them we could see that there were at least four and maybe more. It was also very clear that something important was going on because there were innumerable breaches, fin slaps and tail lobs, even a head butt from the side by one whale on another. Though I had seen two male humpbacks fight before this was a brawl! There was also a lot of trumpeting (a sound males make when antagonized and/or communicating aggression) that I had never heard before...and "blood in the waters".

We had come upon a competitive group of males in full combat and believe me it was a sight to see from close up. As I began taking fluke shots (there were tails everywhere) we noticed that one whale did not enter into the fray. "She" was in the lead, cruising at leisure while the males fought behind her. This was indeed a classic competitive group and I felt gifted to witness it. Humpback whales, as all earth's creatures, vie for genetic survival through some form of selection process. In the case of the great whales it is physical combat (rarely if ever to the death) with the individual having the greatest stamina, strength and intelligence becoming the sole escort and eventual reproductive male. I got six ID's out of that group and a new appreciation for "The Whale People".

Big day in the water with the "critters", 2008

Having made my first swim/dive with a whale the year before I took every opportunity to do so again. My first responsibility was to ID them but then my personal and more intimate journey with them could follow. On April 7th I went out with my friends Shawn and Shawna for an ID run. Within a half hour we came across two humpbacks and after taking my shots we dove in and tried for an encounter. No luck but were within 50' of their "space". Around Punta Pescadero we came across two blue whales (numbers 12 & 13 for the season, very good) and Shawn and I dove in again. We didn't get close but Vicente said that from the panga he could see them go right underneath us. Visibility is always a factor in The Sea of Cortez so they could have been very close without Shawn and I seeing them. This was my first blue whale dive!

After the swim with the blues we went north and found my good friend Alan Pomeroy near San Isidro kayaking with two humpbacks. We all dove in and tried once again for a close encounter but no luck. Nevertheless, what a day in whale world!

Eyeball to eyeball with a blue whale, 2009

In 2009 I invited two interns to join me for that whale season. I had not had any interns down since 2005 so I was ready for some company. Elizabeth Plumb and Jessica Pletz joined me for that year. Late in the season after Jessica had left I went out with Elizabeth and my friend Alan on a search. Unexpectedly we came across a lone blue whale cruising on a very tranquil sea. I got my ID and then suggested we all go in and get "up close and personal" with this critter. Vicente positioned us in front of the whale and as it approached Elizabeth took the best intersect and just as the whale dove, Elizabeth could not have been further away then 40', she dove too. I was right behind her but too far away to see the whale under water. Elizabeth on the other hand got a great view of the blue whale descending into the deep. Elizabeth came up giving a victory call and I thought for sure she was going to swallow a bunch of water and drown. I swam over, hugged her (I was delighted for her) and it was clear she had seen this beautiful creature, up close and personal! I was envious for sure though. She had been close enough to actually see the whale underneath the surface and diving deep into the Sea. My time was to come.

On April 16th of that same year I was blessed with a very special treat. I had taken out my friend Susan Janssen and a friend for a search. It was late in the season but I could afford one more day in the panga so we went out and encountered HB# 42 09 (humpback number 42 for the year 2009) near Boca del Alamo. We then continued to Punta Perico a bit further north to see if any blue whales were coming south through the channel between the mainland and Isla Cerralvo. And there it was, a big, slowly swimming blue whale. I got my two flank shots and asked Vicente to get me in position to swim with it. He did a masterful job and there I was in the water with a blue whale bearing down on me. I was in perfect position, it was going to dive just before it got to me and I could dive right along side it. I waited until the whale's back began to rise higher, and higher, and higher (damn it was big) and then just before getting to me it dove - what a sight! My dive was perfect because there it was, right near me, looking me directly in the eye with what I can only describe as a look of intelligent and incredulous surprise. It rolled to one side to get a longer look, and kept me in sight as it rose for another breath. As it's body lifted to the surface so did I. I felt that I had literally looked into the "eye of the whale" and what traveled between us had been "communion" between two highly (excuse my hubris) intelligent beings.

Swimming with a whale shark, 2010

One day in March we were returning from the south end of the grid and about eight miles out from Punta Pescadero when Vicente spotted a dorsal fin in the water some 50' to our right. I looked for a moment and saw it too but didn't recognize what critter it belonged to. Vicente however said with total assurance, "whale shark, Herme" (he just can't get a handle on Urmas). He sounded so certain of himself and since I had wanted to swim with a whale shark ever since I read about them I donned my ever-ready dive gear and was overboard within a minute. Vicente pointed to where the shark was and I could see his dorsal fin cutting the surface about 20' away.

Whale sharks are real sharks not whales. They are the biggest shark in our oceans today. They are also totally benign and without teeth, they are just way big (up to 45' occasionally). This one was cruising just below the surface so I began to leisurely swim closer to him. He seemed curious but not overly and did not appear at all concerned about my close company. He was moving very slowly and dove a bit deeper for a moment and then began to rise again. As he did I realized he was going to come up right next to me...I mean right next to me. It was then I understood how big he really was. This one was a good 30-35' in length and massively built. As his dorsal fin slowly came up just to the side of me I could not pass up the temptation to grab on to the fin and be taken for a ride. Which is exactly what I did. It was fantastic, a dream, an insane dream. I was doing it, being pulled along by a massive shark (I repeat totally harmless, well...as harmless as any creature can be of that size and strength) and all was ok. Before I go on, I know it was risky and unethical, my only excuse is that I am human, and I really didn't figure this moment of bliss would alter The Universe all that much. At any rate he apparently was not interested in continuing to give me a ride and I felt a quiver in his body, looked behind me and saw his tail (about 8' and terribly powerful) coming around as if to "flick" me off. I kicked (not it) away from it as the tailed passed closer than I would have certainly planned. We did however continue swimming together for a few more minutes, once coming nose to nose for a close look and then he dove and was gone.

Swimming with a pod of Orcas, 2011

By the 2011 season I had swum with humpbacks, blues, and Bryde's whales, a whale shark, a white tipped reef shark, three different species of dolphin, Mobulas (a ray related to the Giant Manta), cow nosed rays, a spotted eagle ray, sea lions and turtles...all "up close and personal". My most heart-pounding (not fearful) swim was yet to come.

On Jan. 31st of 2011 I was taking my neighbors two daughters, Lexie and Tyler out to search for whales. It was still very early in the season but there is always the chance of picking up some ID's nonetheless. One of Vicente's friends radioed us and said he had seen a humpback mother and calf near La Ribera (farthest southern point of my grid). We hustled down there but she was gone when we got into position. A bit disappointed but we had the whole trip back to spot another whale or two.

Around Punta Pescadero Vicente spotted some blows north of us. They were good sized blows but too small for humpbacks. As we closed on them Vicente said, "Orcas"! This got us all excited because you just don't get to see many Orcas even if one spends the amount of time I do out at sea searching. As we closed on them we could see that it was a good sized pod of about ten or twelve with mostly females, some calves and one male. There was lots of thrashing around and when we got near the pod we could see that they were hunting Mobulas. The Mobulas took advantage of our panga and hid underneath, every now and again darting out when it got too crowded only to be grabbed by an Orca. We were all taking pictures and then it crossed my mind that I had always wanted to dive with Orcas (not sure as to exactly why but I can say without hesitation that it has nothing to do with "machisimo") and here was my chance. I hesitated until Tyler said in a soft but clear voice, "isn't this what you want, Urmas?" That's all I needed, I asked Vicente to get me nearer to them and by the time I had my gear on, he had us in the very middle of the hunting group. I dropped over the side and swam into the melee. There were Mobulas all around me and then a fast, dark Orca would pass by, giving me an inquisitive look.

At one point I dove just as an Orca had captured a Mobula in it's mouth and suddenly saw me no more than 10' away. The look she gave me was so comical because she literally spit the Mobula out of her mouth, tilted her head and gave me a "bug eyed" stare that made me laugh.

It was then that I decided that this was my opportunity to play the part of Farley Mowat in the true life movie, *"Never Cry Wolf"*. Instead of helping the wolf pack hunt caribou, I was going to help the Orcas hunt Mobula. I dove under the boat scattering all the Mobulas right into the Orca pod. Then I began swimming around the Mobulas to aid in the hunt. As I did this I noticed that there was no blood in the water or animal parts. What was going on? Then three females swam by me in close formation with four very young Orcas. They were in tight formation and would haven given the Blue Angels a lesson in close flight technique - and then it dawned on me. This was a training session for the calves to learn technique, not gorge themselves on Mobula parts. As the group swept past me I had the funniest feeling that the mothers had brought them close to me in order to make sure they knew I was not part of the menu.

When they moved on I had Vicente pick me up. He asked if I wanted to go in again and I said absolutely! Back in the middle of the pod again I felt like part of the family and was visited by a number of The Orcas as they continued their activities. It had been a long time since I had been so elated. Truly a swim of a lifetime.

Three close encounters in my kayak in one day, 2011

By the first of April of the 2011 season my research funds had run out but there were still whales in the area. It had been a good year with over 40 ID's and a number of good swims. My faith in the quest to see into the eye of the whale had been strongly reinforced by a mother and calf pair that chose to swim with me earlier in the season (see chapter VII). I went out to The WZ (whale zone) in my kayak and settled in to some meditation and sun. The sea was flat, not a cloud in the sky, water was warm and I was feeling good. Within a short time after closing my eyes and beginning my meditation I heard the distinct sound of a whale blow to my north. It was two humpbacks on their way south to take the corner at Cabo San Lucas and begin the long swim north. I paddled towards them but they were too far away to intercept so I just watched them as they drew parallel to me. Then to my surprise, one broke off and headed straight for me. This was a 90 degree change in direction so it was indeed coming to see me. As it came closer I couldn't decide whether to get in the water or not. If you commit to getting in too quickly and they change direction you usually loose them, whereas if you stay in your kayak you can paddle to correct your intercept line. I waited too long, yes it was making directly for me and before I could get in, the whale dove and passed underneath my kayak no more than 10' below me. The water was very clear and I had the opportunity to see it clearly in all it's magnificent detail. It too was getting a good look at me by turning on its side so we were eyeballing each other.

Ah, that was good. How nice that it came over to say hello. Believe me when I say that each and every one of these types of encounters thrills me as much as the first... even more actually.

I settled back into my meditation but before I could get started I see another whale coming from a distance away and from the north again. Well, this is nice I thought, an opportunity for a second encounter. I paddled to get in position but no need, this one also turns and heads straight for me. This is not at all like my experience in 2007 (see above: "Perseverance Furthers") when I was chasing whales all over the bay for an encounter - this time they were coming to me!! This time I did not want to ponder what I will do, I'll let the whale decide. This one again heads directly for my broadside, dives just before reaching me and does the same maneuver, right under the kayak, 10' or so, rolls to the side for an eyeball-to-eyeball look - fantastic! I am really elated, two close encounters in fifteen minutes, and initiated by the whales!!

Now comes the "kicker", yes a third whale breaks off from a new group of three that have now shown up closer to shore and are sure to pass before I can get to them. The one that turned and began towards me I decided was going to be a real close encounter because I just knew it was coming to see me. I dove out of my kayak and waited and there it was, diving just before reaching me (say 30') and again going under me but this time we were no more than 8' apart and I was in the water too. Eyeball to eyeball just doesn't quite cut it! Same roll to the side, same benevolent and inquisitive look, and the same heart warming feeling of acceptance by "The Whale People"! I literally could not believe my good fortune. These were not chance encounters, nor encounters forced by me, these were completely and totally initiated by the whales themselves.

A memorable swim with sperm whales, 2012

The 2012 season had just begun. I had eleven ID's so far but I was yet to swim with them. I was getting older (70) and the water seemed colder at this time of year, but my gear is always with me, including the items I need to make a rescue, because I never know when that might be required. As it turned out the season was not as plentiful as previous seasons, I only acquired twenty three IDs but

I had a number of very memorable swimming encounters with my whale friends.

On April 11th I took some friends with me as spotters on a run through my grid. We came upon a lone humpback and it became HB# 22 12. Later in the morning we spotted a number of good sized blows further out in the sea. It was difficult to tell just what we had because there were so many yet too big to be dolphins. It was also hard to believe they could be humpbacks (too many and the blow was not right) nor blues (the blow was not again not right), yet they did not behave like Orcas either.

When we got closer we discovered that they were sperm whales, rare in The Sea of Cortez but not unique. The only other time I had seen sperm whales was in 2007, and though captivated, I was not ready to swim with a full on "Moby Dick"! Later I vowed I would, well…later had arrived! This time however instead of four there were twenty to twenty five. After taking a few token camera shots (I'm not ID'ing sperm whales) I realized that this was a chance of a lifetime, so without even putting on my wet suit or fins I bailed out of the panga and began swimming with the whales. Vicente said later that I was swimming along side them as close as 20'. It was an incredible high, I was part of a herd of sperm whales! They were swimming in all directions and seemed to be totally at ease with my presence…as I was with theirs; I could not have felt safer. I feel that this was a gift from "The Whale People", a wonderful new level of interaction/communion. What might this portend for the 2013 season?

Last swim with a humpback for the 2012 season?

In mid July of 2012 I had a very unexpected and incredible day in The WZ. I went out early and was kayaking out to the reef when I caught some movement further out... panga, no, a lone humpback!!! It was moving south and about a half mile out from me. The sea was absolutely flat, no boats, just me and the whale, I mean we were alone. I just knew I could intercept him. I paddled like hell and on it's next blow series I was within 200'. I changed my angle just a bit and continued paddling. On the next blow series, the whale surfaced 50' to my left. He fluked up after only two blows and I continued in the direction he was going at a moderate paddle. Then he came up again about 40' behind me and coming directly for me. It was too late to dive in but within a moment I could see him perfectly under the water just to my right at about 10' depth and 15' away...beautiful beyond description. The entire body so close, so clear, so intelligent and so very alive. I kayaked above him, just to his left until he came up for his next blow and he was right next to me! I could have touched him with my paddle. He fluked up and I again kayaked in the direction I thought was right. After about five minutes another blow, again right behind me. I grabbed my mask, rolled in and waited. Then he came...right to me!! He rolled a little to his left side, took a good look at me, I waved and dove to him. He continued on his way and I swam behind him till he fluked once more. Too tired to follow further, but this was my longest, clearest, closest humpback encounter so far! hoohah!!!! I feel deeply that this was a gift from

The Great Mystery and The Whale People; I would like to think I have earned it!

I purposely do not carry a camera when I kayak because I do not want the lens to get in the way of the experience. I have missed some extraordinary shots but I am absolutely certain it is the right decision.

I am confidant that I will have many more close encounters in the years to come. And if I continue to be blessed by The Great Mystery with good health - I have more projects in mind!

Appendix B
A Short Biography

I was born in 1941 in Tallinn, Estonia. The first three years of my life were spent in an atmosphere created by the violence of World War II. After the deportation of my father, uncle and grandfather to Siberia it was decided that we would attempt an escape across the Baltic Sea. This was successful and resulted in our eventual emigration to America and ultimately to the state of California.

My first memories are of Redwood City in 1947 and in particular my first Halloween. Not aware of the tradition I was both frightened and then isolated by the fact that I had no costume. I found America bewildering but not unmanageable. As time went by I found that I was a natural athlete and due to this was accepted into the ranks of the male elite by the fourth grade. Because of those abilities, particularly in swimming and water polo, I continued to have social success in high school. I also discovered during this time an interest in science for it provided some answers to a world which was otherwise difficult to understand and at various times, to enjoy.

Biology, specifically the animal world and the physiochemical workings of the single cell, was the academic area I was intrigued by most. By the time I began college at Washington State University in 1960, history and anthropology also began to occupy my interest. I transferred to San Jose State College in 1961 in order to get away from the cold and drab eastern Washington plains. There I studied zoology and was fortunate to have had some very excellent professors who stoked the fires of my curiosity and gave me a lifelong appreciation of good teaching. I graduated in February of 1965 and in order to broaden my experience I married and traveled to Europe for an extended leave. While there I worked as a Practicant Microbiologist at Bjare Industries in Sweden, a laborer in Berchtesgaden, Germany, and an apprentice seaman on my return to the states. I drove across the country from Connecticut with my wife and decided to continue my education at the University of Arizona in Medical Microbiology. One of my dreams had been to be an MD traveling all over the world on the ship *Hope* saving the lives of millions in The third World. Not being accepted to medical school made that impossible so I chose a field closely aligned to medicine. My time was spent attempting to discover the role of blood complement in the immune response while determining where I stood in regard to The Viet Nam War, the environment and human rights. I was able to meld these pursuits successfully until my Masters Degree had been conferred in 1968 and I began my Ph.D. work. After two more years at the U of A I could no longer maintain interest in the field, and returned to California to work as a truck driver and welder for the PG&E gas division. My first child, Kersti, was born during this time and I delighted in her presence.

In 1970 I obtained a job at the UC Berkeley Marine Lab in Bodega Bay, where I became their collector and boat handler. My second child, Zackary, was born during this time and I was blessed with a fine little boy. I worked there two and a half years and then returned to Palo Alto to engage in an apprenticeship in furniture finishing and restoration while also developing a permanent outdoor sign for Permaloy Corp. of Utah and writing EIR's for my brothers engineering company. In 1973 I was hired as the first biology and chemistry instructor at Mendocino Community College in Ukiah. I taught there full time for two years and was a part time Professor until 2009. Over the years I have expanded my teaching into every academic venue (public school 5-12, Waldorf School teacher and board chairman, Janus senior citizens program, ElderHostel director, curriculum advisor for D-Q University (The Native American University outside of Sacramento), instructor on the Covelo Indian Reservation and operated a private mentoring service in the county. In 1990 I returned to school and completed a Ph.D. at the Western Inst. for Social Research in Berkeley. My doctoral dissertation, *Education as a Ritual Process*, gave me the opportunity to organize and articulate my 25 years of teaching discovery and experience.

For five years I was the executive director of Pelagikos: World Marine Research, which has taken me to sea in the pursuit of knowledge pertaining to the whales and dolphins. I also sat on the board of the Cetacean Studies Institute of Santa Fe, NM and the Cloud Forest Inst. in Equador. In 1998 I spent 6 months on board the 84' sailing schooner Dariabar as a crewman. We sailed from San Francisco to Cabo San Lucas, Mexico and then to Hawai'i. We were part of the ATOC experiment, specifically to monitor humpback whale activity during the tests. In the year 2000 I traveled to Egypt in order to better understand the grandeur of what once was. I was of course duly impressed. Since then I have been engaged in a photo-identification project involving humpback whales in the eastern pacific. I am a collaborator with Dr. Jorge Urban ramirez of The University in La Paz, Mexico and with Richard Sears the Director of The Mingan Island Whale Project in Canada.

Now that I am on my way to my 71st year of life, I am confident that I have acquired the necessary knowledge and experience to direct my energies to where they might do the most good. Although it is undoubtedly in education that I am strongest, I wish to broaden my audience to as many people as I can. We are entering into one of the greatest, if not the greatest, paradigm shift in the history of our species. It will be at least a part of my service to provide information to those who are concerned as to what the shift might imply in some areas. As the Hopi Elders say, "it is the last of the eleventh hour, yet it could be a very good time".

Urmas Kaldveer, Ph.D.
Executive Director and "Whale Talker"
Mendocino Institute of Science & History

Appendix C
The Humpback Whale

The humpback whale (Megaptera novaeangliae) is a species of baleen whale. One of the larger rorqual species, adults range in length from 12–16 metres (39–52 ft) and weigh approximately 36,000 kilograms (79,000 lb). The humpback has a distinctive body shape, with unusually long pectoral fins and a knobbly head. It is an acrobatic animal, often breaching and slapping the water. Males produce a complex song, which lasts for 10 to 20 minutes and is repeated for hours at a time. The purpose of the song is not yet clear, although it appears to have a role in mating.

Found in oceans and seas around the world, humpback whales typically migrate up to 8,000 miles or more each year. Humpbacks feed in summer in polar waters and migrate to tropical or sub-tropical waters to breed and give birth in the winter. The species' diet consists mostly of krill and small fish. Humpbacks have a diverse repertoire of feeding methods, including the bubble net feeding technique.

Like other large whales, the humpback was and is a target for the whaling industry. Due to over-hunting, its population fell by an estimated 90% before a whaling moratorium was introduced in 1966. Stocks have since partially recovered; however, entanglement in fishing gear, collisions with ships, and noise pollution also remain concerns. There are at least 80,000 humpback whales worldwide. Once hunted to the brink of extinction, humpbacks are now sought by whale-watchers, particularly off parts of Australia, New Zealand, South America, Canada, and the United States.

Humpback whales are rorquals (family Balaenopteridae), a family that includes the blue whale, the fin whale, the Bryde's whale, the sei whale and the minke whale. The rorquals are believed to have diverged from the other families of the suborder Mysticeti as long ago as 3 million years.

Though clearly related to the giant whales of the genus Balaenoptera, the humpback has been the sole member of its genus since Gray's work in 1846. More recently though, DNA sequencing analysis has indicated the Humpback is more closely related to certain rorquals, particularly the fin whale (Balaenoptera physalus), and possibly to the gray whale (Eschrichtius robustus), than it is to rorquals such as the minke whales. If further research confirms these relationships, it will be necessary to reclassify the rorquals.

The humpback whale was first identified as "baleine de la Nouvelle Angleterre" by Mathurin Jacques Brisson in his Regnum Animale of 1756. In 1781, Georg Heinrich Borowski described the species, converting Brisson's name to its Latin equivalent, Balaena novaeangliae. Early in the 19th century Lacépède shifted the humpback from the Balaenidae family, renaming it Balaenoptera jubartes. In 1846, John Edward Gray created the genus Megaptera, classifying the humpback as Megaptera longpinna, but in 1932, Remington Kellogg reverted the species names to use Borowski's novaeangliae. The common name is derived from the curving of their back when diving. The generic name Megaptera from the Greek mega-/"giant" and ptera/"wing", refers to their large front flippers. The specific name means "New Englander" and was probably given by Brisson due the regular sightings of humpbacks off the coast of New England.

A humpback whale can easily be identified by its stocky body with an obvious hump and black dorsal coloring. The head and lower jaw are covered with knobs called tubercles, which are actually hair follicles, and are characteristic of the species. The fluked tail, which it lifts above the surface in some dive sequences, has wavy trailing edges. The four global populations, all under study, are: North Pacific, Atlantic, and Southern Ocean humpbacks, which have distinct populations which complete a migratory round-trip each year and the Indian Ocean population, which does not migrate, prevented by that ocean's northern coastline.

The long black and white tail fin, which can be up to a third of body length, and the pectoral fins have unique patterns, which make individual whales identifiable. Several hypotheses attempt to explain the humpback's pectoral fins, which are proportionally the longest fins of any Cetacean. The two most enduring mention the higher maneuverability afforded by long fins, and the usefulness of the increased surface area for temperature control when migrating between warm and cold climates. Humpbacks also have 'rete mirable', a heat exchanging system, which works similarly in humpbacks, sharks and other fish.[citation needed]

Humpbacks have 270 to 400 darkly coloured baleen plates on each side of the mouth. The plates measure from a mere 18 inches (46 cm) in the front to approximately 3 feet (0.91 m) long in the back, behind the hinge. Ventral grooves run from the lower jaw to the umbilicus about halfway along the underside of the whale. These grooves are less numerous (usually 16–20) and consequently more prominent than in other rorquals.

The stubby dorsal fin is visible soon after the blow when the whale surfaces, but disappears by the time the flukes emerge. Humpbacks have a 3 metres (9.8 ft) heart-shaped to bushy blow, or exhalation of water through the blowholes. Because Humpback Whales breathe voluntarily, researchers have said that it is possible that the whales shut off only half of the brain when sleeping.

Newborn calves are roughly the length of their mother's head. At birth, calves measure 15-20 feet (6.1 m) at 2 short tons (1.8 t) The mother, by comparison, is about 50 feet (15 m). They nurse for approximately six months, then mix nursing and independent feeding for possibly six months more. Humpback milk is 50% fat and pink in color. Some calves have been observed alone after arrival in Alaskan waters.

Females reach sexual maturity at the age of five, achieving full adult size a little later. Males reach sexual maturity at approximately 7 years of age. The humpback whale lifespan ranges from 45–100 years.

Fully grown, the males average 15–16 metres (49–52 ft). Females are slightly larger at 16–17 metres (52–56 ft), and 40,000 kilograms (44 short tons); the largest recorded specimen was 19 metres (62 ft) long and had pectoral fins measuring 6 metres (20 ft) each.

Females have a hemispherical lobe about 15 centimetres (5.9 in) in diameter in their genital region. This visually distinguishes males and females. The male's penis usually remains hidden in the genital slit. Male whales have distinctive scars on heads and bodies, some resulting from battles over females.

Identifying individuals

The varying patterns on the tail flukes are sufficient to identify individuals. Unique visual identification is not currently possible in most cetacean species (other exceptions include orcas and right whales), making the humpback a popular study species. A study using data from 1973 to 1998 on whales in the North Atlantic gave researchers detailed information on gestation times, growth rates, and calving periods, as well as allowing more accurate population predictions by simulating the mark-release-recapture technique (Katona and Beard 1982). A photographic catalogue of all known North Atlantic whales was developed over this period and is currently maintained by College of the Atlantic. Similar photographic identification projects have begun in the North Pacific by SPLASH (Structure of Populations, Levels of Abundance and Status of Humpbacks), and around the world.

Reproduction

Females typically breed every two or three years. The gestation period is 11.5 months, yet some individuals have been known to breed in two consecutive years. The peak months for birth are January, February, July, and August. There is usually a 1-2 year period between humpback births. Humpback whales were thought to live 50–60 years, but new studies using the changes in amino acids behind eye lenses proved another baleen whale, the bowhead, to be 211 years old. This animal was taken by the Inuit off Alaska.

Recent research on humpback mitochondrial DNA reveals that groups that live in proximity to each other may represent distinct breeding pools.

Social structure

The humpback social structure is loose-knit. Typically, individuals live alone or in small, transient groups that disband after a few hours. These whales are not excessively social in most cases. Groups may stay together a little longer in summer to forage and feed cooperatively. Longer-term relationships between pairs or small groups, lasting months or even years, have rarely been observed. It is possible that some females retain bonds created via cooperative feeding for a lifetime. The humpback's range overlaps considerably with other whale and dolphin species—for instance, the minke whale. However, humpbacks rarely interact socially with them, though humpback calves in Hawaiian waters sometimes play with bottlenose dolphin calves.

Courtship

Courtship rituals take place during the winter months, following migration toward the equator from summer feeding grounds closer to the poles. Competition is usually fierce, and unrelated males dubbed escorts by researcher Louis Herman frequently trail females as well as mother-calf dyads. Groups of two to twenty males gather around a single female and exhibit a variety of behaviors over several hours to establish dominance of what is known as a competitive group. Group size ebbs and flows as unsuccessful males retreat and others arrive to try their luck. Behaviors include breaching, spyhopping, lob-tailing, tail-slapping, fin-slapping, peduncle throws, charging and parrying. Less common "super pods" may number more than 40 males, all vying for the same female. (M. Ferrari et al.)

Whale song is assumed to have an important role in mate selection; however, scientists remain unsure whether song is used between males to establish identity and dominance, between a male and a female as a mating call, or both.

Song

Both male and female humpback whales vocalize, however only males produce the long, loud, complex "songs" for which the species is famous. Each song consists of several sounds in a low register that vary in amplitude and frequency, and typically lasts from 10 to 20 minutes. Humpbacks may sing continuously for more than 24 hours. Cetaceans have no vocal cords, so whales generate their song by forcing air through their massive nasal cavities.

Whales within a large area sing the same song. All North Atlantic humpbacks sing the same song, and those of the North Pacific sing a different song. Each population's song changes slowly over a period of years without repeating.

Scientists are unsure of the purpose of whale song. Only males sing, suggesting that one purpose is to attract females. However, many of the whales observed to approach a singer are other males, and results in conflict. Singing may therefore be a challenge to other males. Some scientists have hypothesized that the song may serve an echolocative function.[17] During the feeding season, humpbacks make altogether different vocalizations for herding fish into their bubble nets.

All these behaviors also occur absent potential mates. This indicates that they are probably a more general communication tool. Scientists hypothesize that singing may keep migrating populations connected. (Ferrari,

Nicklin, Darling, et al.) Some observers report that singing begins when competition for a female ends.

Humpback whales have also been found to make a range of other social sounds to communicate such as "grunts", "groans", "thwops", "snorts" and "barks".

Ecology

Humpbacks feed primarily in summer and live off fat reserves during winter. They feed only rarely and opportunistically in their wintering waters (this has proved not to be true). The humpback is an energetic hunter, taking krill and small schooling fish such as Atlantic herring, Atlantic salmon, capelin, and American sand lance as well as Atlantic mackerel, pollock, and haddock in the North Atlantic. Krill and copepods have been recorded as prey species in Australian and Antarctic waters Humpbacks hunt by direct attack or by stunning prey by hitting the water with pectoral fins or flukes.

The humpback has the most diverse feeding repertoire of all baleen whales. Its most inventive technique is known as bubble net feeding: a group of whales swims in a shrinking circle blowing bubbles below a school of prey. The shrinking ring of bubbles encircles the school and confines it in an ever-smaller cylinder. This ring can begin at up to 30 metres (98 ft) in diameter and involve the cooperation of a dozen animals. Using a crittercam attached to a whale's back it was discovered that some whales blow the bubbles, some dive deeper to drive fish toward the surface, and others herd prey into the net by vocalizing. The whales then suddenly swim upward through the 'net', mouths agape, swallowing thousands of fish in one gulp. Plated grooves in the whale's mouth allow the creature to easily drain all the water that was initially taken in. Solitary humpbacks have also been observed employing this technique.

Predation

Given scarring records, killer whales are known to prey upon juvenile humpbacks and this has now been recorded on film. The result of these attacks is generally nothing more serious than some scarring of the skin, but it is likely that young calves are often killed.

Range and habitat

Humpbacks inhabit all major oceans, in a wide band running from the Antarctic ice edge to 77° N latitude, though not in the eastern Mediterranean or the Baltic Sea.

Humpbacks are migratory, spending summers in cooler, high-latitude waters and mating and calving in tropical and subtropical waters. An exception to this rule is a population in the Arabian Sea, which remains in these tropical waters year-round. Annual migrations of up to 25,000 kilometres (16,000 mi) are typical, making it one of the mammal's best-traveled species.

A large population spreads across the Hawaiian islands every winter, ranging from the island of Hawaii in the south to Kure Atoll in the north. A 2007 study identified seven individuals wintering off the Pacific coast of Costa

Rica as having traveled from the Antarctic—around 8,300 kilometres (5,200 mi). Identified by their unique tail patterns, these animals made the longest documented mammalian migration.

In Australia, two main migratory populations have been identified, off the west and east coast respectively. These two populations are distinct, with only a few females in each generation crossing between the two groups.

This article comes from The Free Wikipedia Commons

Appendix D
Aboriginal Dreamtime

To the Australian Aboriginals, whose traditional homelands encompass the coastal regions of the Australian continent, the powerful and boisterous whale is a beloved ancestor, shaper of the landscape, and immortal being of that timeless, instructive and never-ending epoch of creation and earthly transformation widely known as the Dreamtime or Dreaming. To the coastal tribes the whale is the all-powerful Rainbow Serpent and is closely associated with the Rainbow Serpent/Snake of the inland. The whale is, as is the serpent elsewhere in the world, associated with fire, earth energy, wind, water, the sun, moon and the symbol that links all of these elements - the rainbow.

In most coastal tribal stories the whale, after arriving from his ancestral home in the Milky Way, causes the seas to rise, brings other creatures with him, and travels into a sea cave home moving inland to emerge via blowholes, caves and waterholes into the sunlight of the inland. To the 'People of the whale', blowholes, caves and waterholes were sacred because they were the aperture through which the whale ancestor made his first appearance on earth. These places were also sacred because the whale's presence continues to animate them, to pulsate from them in ways that challenge the human imagination and that permeate the natural world.

During the course of the whales land sculpting journey on and under the earth's Dreamtime surface, the whale left in its wake a sacred, eternal living essence. Rock formations, (both natural and man-made), trees, waterholes, and other features that dot the local terrain, mark these ancestral Dreamtime passages, record the dramas, and entomb many of its principal characters, as if in sleep. At the same time, these sacred places are centers of Natures reproductive powers.

The whale's story is a reflection of the whale 'totem' people's Dreamtime origins, religious duties, and of the primal, cyclic, life-perpetuating processes of the natural world. On at least one level the Whale embodies the essence of nature's life force and fertility, in particular the fertility of the waters. Consequently the items used to harvest the produce of the sea; spears, nets, baskets, and stone fish traps, are associated with and are sacred aspects of the whale mythology.

The whale clans knowledge that the local terrain is a sacred map of the whale's ancestral journey gives them an extraordinary sense of participation in the workings of local ecosystems. The initiated members of the clans could communicate directly through ritual and 'prayer', with the forces of Nature. Radiating out from the sacred whale dreaming places along the Dreamtime trails of the ancestral whale are indelible geographic points, where the aboriginals understanding of the Creation Time connection between the land and nature's mysterious regenerative powers allows them to work in harmony with the forces of fertility at these locations.

During the whale's treks across the landscape, he left a trail of words and musical notes as well as indelible physical footprints, permanently etching the whale's story in the earth. Generations later, by devoutly singing the appropriate sacred song, a totemic clansman could reliably navigate for days along these great Dreamtime tracks. In the process, the land would be rejuvenated, memories of totemic ancestors would be rekindled, and reunions with faraway kinsmen would take place along the ancient Dreamtime trails of stone, story and song. In one sense the whole of Australia could be read as a musical score of songlines. The songlines could, in a sense, be visualized as a serpent, writhing this way and that, in which every 'episode' was readable in terms of geology and natural phenomena.

Beyond lending a sense of spirituality and cosmic order within the aboriginal world, these ancient beliefs convey genuine ecological insights into the working of Nature. Stories of the Dreamtime travels of the whale reveal a sophisticated grasp of whale ecology. Maps of the whale's journeys, breathing life and form into the landscape as he went, correspond with uncanny precision to maps of the preferred habitats of the whale that have recently been painstakingly assembled by detailed aerial and ground based scientific study.

Traditional aboriginal beliefs about the sacred sites of the Whale Dreaming represent a remarkable fusion of ecological and spiritual knowledge. They encode genuine ecological truths about the population dynamics and location data, with sacred places corresponding to prime whale breeding habitat and with places where whales 'like to linger'. The Mirning tribe's names for the Southern Right Whale, Numbadda, means "to hang around wanting something", reflecting this species tendency to favour particular locations to mate, give birth and rear calves, or just linger close to shore. At the same time, unlike sterile scientific findings, they contain a moral code mandating irrevocable human responsibility to honour and nurture the precious, life-sustaining whale populations and nature in general.

Appendix E

The Legend of the Golden Dolphin

A Synopsis

The Legend of the Golden Dolphin is a collection of stories. The stories are factual; the stories are mythical; the stories are fantastic; the stories are religious; the stories are historical. The stories tell of the beginnings of human history and the inter-related tale of the beginnings of Cetacean/Human contact and leads to the subtle and profound effects of that contact. The Legend also foretells a possible future for us all. In its intent, the Legend is a key to the Human dilemma, the challenge that we face in attempting to become all that we are capable of becoming. It is about Humans and the sister race of the Dolphins.

The Legend is not concerned with presenting itself as a 'revealed truth', as the 'new dogma'. It is concerned with the stimulation of the human heart, thru the agency of the mind, to re-awaken the truth of the inter-connectedness of all life. Truth, to the heart, is the feeling of "Yes". The Legend awakens, mystifies, amuses and astonishes the mind, and the heart says

"Yes". Myths and legends can be defined as truths that cannot be adequately contained by measureable 'facts', and it is in this definition that we can begin to grasp the intent of the Legend.

Following many paths, the Legend reveals the beginnings of human civilization. It reflects on the question of how humans came to make such startling forward leaps in becoming civilized. How did the stone-tool-using peoples of the Nile River become, apparently in a period covering only 200 years, an advanced civilization with highly developed architecture, art, writing, law, agriculture etc.? What was the secret of the Sumerian cultural ascendancy? Who were the amphibious beings described by the ancient peoples as the "Givers of the Civilizing Arts"? Who was Oannes, or more accurately asked, who were the Oannes (plural)? And what of the first incarnation of Vishnu, as a half-man, half-fish who showed the peoples of ancient India the ways of civilization? How did Jesus become associated with a fish-like symbol with a dolphin's tail? What role did dolphins and whales play in the development of the key ideas upon which the greatest of human cultures were built? The roots of every one of the worlds great cultures and religions intertwine here, where tales of dolphins inspiring humans are to be found.

Mysteries abound in this story. The Dogon people of Mali, who seem to be able to trace themselves back to a homeland near the Mediterranean, have a very sophisticated knowledge of the star Sirius, the brightest star in our sky. Their knowledge of this star includes several very accurate details which have been only recently verified by astronomers. The Dogon tell of having this knowledge from great beings who came to earth far in the past, from a planet which circles Sirius, to give them the arts of civilization. The beings are depicted as huge water creatures who breathe air and have flukes. Part of the tale the Dogon tell is of these beings, the Nommo, who will have one-third of their number sacrificed "on the altar of human greed", yet whose sacrifice will bring about "the purification and reorganization of the Universe". In addition, it is said in the Dogon tradition that "the Nommo will arise in human form and descend upon the Earth".

The Legend, as told to its messenger, says that the reincarnated souls of the dolphins and whales slaughtered over the last 300 years will return to become "The Force of Freedom", inspired by, and under the guidance of, The Golden Dolphin. Today, many of those to whom the Legend has deep significance consider themselves to be, in some mysterious way, part of this movement toward a spiritually, environmentally, and socially sustainable world.

Dolphins and whales are beings of extraordinary qualities, with built-in futuristic technologies, the largest brains on Earth and by far the longest continuous history of sophisticated existence on Earth. While biological creatures living in the world of predator and prey, they communicate, they socialize, they cooperate, they play much of their days, they are gentle, compassionate, intelligent, and loving in the main, and all of the other traits we hold in highest esteem. They have proven telepathic abilities; they have demonstrated healing effects; they carry their enthusiasm for Life into the lives of nearly all humans who encounter them. In short, they are spiritually advanced beings who are in fact a sister race, awaiting us on the other side of the portal of 'planetary initiation' that the human race is rapidly approaching.

According to the Legend, as Masters of Sound, the Cetaceans await the human family, the yet-to-be-fully-realised Masters of Light. Together, the two families, who are in fact one, will proceed into Universe as realized beings, joining the 'Family of the Stars'.

The Legend of the Golden Dolphin was first encountered by an Australian named Peter Shenstone. In meditation one night, he suddenly found himself immersed in an oceanic sound that resolved itself into meaningful information. Peter was the recipient of this tale over a period of almost two weeks, each night filled with huge amounts of information and clues pouring in. During the next several years, Peter and his wife Jan, and several friends, began collecting the elements of the story from sources far and wide. In its physical form, it resides in a box, collected into about ten large-format volumes of drawings and text, assembled in a graphically beautiful way. The Legend is conveyed to listeners as an oral telling, a classic story-telling, with the presenter turning the pages of the volumes, revealing the images, ideas, and intent of the Golden Dolphin in a fluid and gracious manner. The telling can take from four hours to several days, depending on the receptivity of the audience, the inspiration of the teller, and the needs of the moment.

The Legend of the Golden Dolphin, in its box, has traveled to America several times. Each time, a potent wave of awakening was created, spreading out into the rest of the world. The Dolphins asked Peter to share the dream, tell the tale, and invite those who listen to join him in the dreaming, and work, indicated by the tale. Since encountering the Legend in 1983, it became my own mission to understand it, help to develop it further, and to share it widely. It has been shared around the world, in thousands of illustrated presentations, to tens of thousands of listeners. And, the Legend of the Golden Dolphin remains a mystery, a tale intended to illuminate, inspire, and activate us, to become like dolphins in some ways. Rising to another dimension of life, they draw inspiration from the world of air and land, and we can rise too, into realms of intellectual and spiritual light, to become 'enlightened', caring, co-evolved partners, enjoying the Earth and sharing it equally.

The central message of the Legend of the Golden Dolphin is not made explicit. It is, instead, left to each person who experiences it, to find its meaning for them, to respond as an individual. The Legend has been described as being like the game of 'tag': once it has touched you, you are 'it'! How you will work out your own relationship to its message is up to you. In the end, it takes a spiral path, directly into your heart.

By C. Scott Taylor 1995

Revised, 2012

Epilogue

In November of 2012 the whales will begin to reappear again in our waters here in the east cape. By February and March of 2013 they will be abundant enough for me to warrant the use of my research funds to search for and photo-ID them. Each year is different and each year offers new opportunities to engage with them both scientifically and personally. I will do all I can to expand my relationship with them and to perhaps, if they so desire, to be in communion with them. Their presence in my life has been an utter joy and has gone a long way towards helping heal my inner child.

Perhaps *The Others* will compel you to reevaluate your concept of the sea and it's critters. Perhaps you will choose to become part of The Delphic Wave. I hope so! The need for the study and the protection of the great whales has only begun. Please consider a donation to my work. If my whale dream continues, we may one day have a small marine lab attached to our local school here in El Cardonal where the children can learn about the wonder that is right before them—maybe one will become a "whale talker".

More information is available on my website: www.urmas-kaldveer.com

I also maintain a blogspot where it is possible to keep up with my activities monthly: www.urmkal.blogspot.com

Donations towards my whale work are always welcome and needed. The Mendocino Institute of Science and History is a 501(c)(3) non-profit corporation in California, all donations are tax deductible.

Checks are payable to: MioSah
                             C/O Susan Janssen
                             106 Canyon Dr.
                             Ukiah, CA 95482

CPSIA information can be obtained at www.ICGtesting.com
Printed in the USA
LVOW12s0337201213

365886LV00001B/49/P